THE SPACE BARONS

THE
SPACE
BARONS

ELON MUSK, JEFF BEZOS, AND THE QUEST TO COLONIZE THE COSMOS

Christian Davenport

PUBLICAFFAIRS
New York

PublicAffairs
Hachette Book Group
1290 Avenue of the Americas, New York, NY 10104
www.publicaffairsbooks.com
@Public_Affairs

Printed in the United States of America
First Edition: March 2018

Published by PublicAffairs, an imprint of Perseus Books, LLC, a subsidiary of Hachette Book Group, Inc. The PublicAffairs name and logo is a trademark of Hachette Book Group.

The Hachette Speakers Bureau provides a wide range of authors for speaking events. To find out more, go to www.hachettespeakersbureau.com or call (866) 376–6591. The publisher is not responsible for websites (or their content) that are not owned by the publisher.

Library of Congress Cataloging-in-Publication Data
Names: Davenport, Christian, author.
Title: The space barons : Elon Musk, Jeff Bezos, and the quest to colonize the cosmos / Christian Davenport.
Description: First edition. | New York : PublicAffairs, [2018] | Includes bibliographical references and index.
Identifiers: LCCN 2017053089| ISBN 9781610398299 (hardcover) | ISBN 9781610398305 (ebook)
Subjects: LCSH: Space industrialization—United States. | Industrialists—United States—Biography. | Aerospace engineers—United States—Biography. | Bezos, Jeffrey. | Musk, Elon. | Blue Origen (Firm) | SpaceX (Firm) | Aerospace industries—United States. | Outer space—Civilian use.
Classification: LCC TL789.85.A1 D38 2018 | DDC 338.7/6294092273 [B]—dc23
LC record available at https://lccn.loc.gov/2017053089

ISBNs: 978-1-61039-829-9 (hardcover); 978-1-61039-830-5 (e-book)

LSC-C

10   9   8

For Heather

CONTENTS

All there is to thinking is seeing something noticeable which makes you see something you weren't noticing which makes you see something that isn't even visible.

—NORMAN MACLEAN,
A RIVER RUNS THROUGH IT

INTRODUCTION

"Touchdown"

T HEY CAUGHT THEIR first glimpse of it at 25,000 feet and falling fast. Normally, a rocket dropping like a bomb would be cause for panic. But instead, the four hundred or so people gathered in the employee lounge at Blue Origin's headquarters outside Seattle were thrilled to see the booster plummeting toward Earth.

"Estimate ten seconds to engine start," the flight controller announced.

The employees, mostly engineers, were packed in, watching the rocket in free fall on a giant screen. Some had their hands over their mouth. Others sat forward with fists clenched. Mostly, they were silent, waiting for what would happen next.

"Engine start," said the flight controller. "We have thrust."

At that, the employees started cheering wildly. Just minutes before on this morning three days before Thanksgiving in 2015, the engine had fired to lift the rocket off the launchpad at Blue Origin's West Texas test site, flying it faster than the speed of sound past the 62-mile threshold that's considered the edge of space. But now that the rocket was falling back, the thrust had the opposite effect: it was slowing the rocket down, preventing it from slamming into the ground and exploding.

Soon the rocket's altitude was 2,000 feet.

Then 1,000.

500.

As the ground came into view, fire from the engine kicked up a plume of dust. The employees at Blue Origin rose to their feet in unison. The rocket was under control, descending gently, like a hot-air balloon coming in for a landing.

"One hundred and fifty feet," the flight controller called out.

"Seventy feet."

"Fifty feet. Velocity steady."

There was one last flash of the engines, a bright orange glow shining through the dust and smoke. Then, it went out.

"Touchdown."

The room broke out in pandemonium. The employees celebrated wildly, hugging one another, giving high fives. The rocket booster stood in the center of the pad like a giant trophy.

Jeff Bezos had watched from the control room of his company's West Texas launch site. It was "one of the greatest moments of my life," he would later say. "I was misty-eyed."

Twenty-eight days later, another rocket was falling from the sky. This time, it was a much bigger booster that had been flying at a much greater velocity, a speed capable of crossing not just the threshold of space but of getting its payload to orbit Earth. For this landing attempt, the chances of success were even more improbable, the chance of disaster, far greater.

About ten minutes after blasting off into the dark, evening skies over Cape Canaveral, Florida, the fire from the rocket engine suddenly appeared like a streetlight in the distance, a shimmering, ethereal beacon lowering through the clouds.

As they watched on television screens, the SpaceX employees who had gathered at the company's headquarters outside Los Angeles on this evening just before Christmas 2015 cheered it just as their rivals at Blue Origin had done—and then some.

Elon Musk watched the rocket reappear from outside on a causeway. Then, he sprinted back into the control room to see the image of the rocket standing proudly on the landing pad. Like Bezos, he would say this was one of the greatest days of his life. A "revolutionary moment," he called it, one that "quite dramatically improves my confidence that a city on Mars is possible."

SOME FIFTY YEARS after the advent of the Space Age, no one ever had flown a rocket past the edge of space and landed it vertically. Now, it had been performed twice in less than a month.

For generations, spaceflight had been celebrated largely for the take-offs. But the landings were reminiscent of Neil Armstrong and Buzz Aldrin's touching down on the surface of the moon in the lunar module. Or the "seven-minutes of terror" landing of the Curiosity rover on the surface of Mars. The sight of the boosters standing on terra firma, scorched but triumphant, portended a sense of arrival, and offered hope for another Apollo 11 moment, the next giant leap many had felt they were promised but had never come.

Even more impressive was that the landings had been accomplished not by nations—not even NASA had pulled off such a feat—but by a pair of private companies. Backed by billionaires intent on developing reusable rockets, which could fly, land, then fly again, they were pursuing a holy grail—a technology with the potential to dramatically lower the cost of space travel.

For decades, the first stages of rockets were ditched into the ocean after powering their payloads to space. To Musk and Bezos, that was an incredible waste, like throwing away an airplane after flying from New York to Los Angeles. Now they had shown that rockets could fly not just up, but back down, landing with precision and reigniting interest in human space travel in a way not seen in decades.

The landings had touched off celebrations not just at Blue Origin and SpaceX, but among their legions of growing fans, who watched the viral videos by the millions. It was the 1960s revisited, but on YouTube and Reddit, where the new space fans congregated the way enthusiasts once crowded Cocoa Beach along the cape. With unbridled enthusiasm, they cheered this new Space Age, just as their parents cheered John Glenn blasting off to orbit in a moment that eroded Walter Cronkite's steely, newsman's detachment. "Oh, go baby!" he had gushed, live on air, as that rocket tore a hole in the sky.

MUSK AND BEZOS were the leaders of this resurrection of the American space program, a pair of billionaires with vastly different styles and temperaments. Always audacious, Musk had plowed far ahead, his triumphs and failures commanding center stage. Bezos remained quiet and clandestine, his mysterious rocket venture kept hidden behind the curtain.

But there were others. Like Bezos, Richard Branson was promising to fly tourists past the edge of space to get glimpses of Earth from above and experience a few minutes of weightlessness. Paul Allen, the cofounder of Microsoft, who had backed the first commercial spacecraft to reach space, was now building the largest airplane the world had ever seen. Bigger than Howard Hughes's Spruce Goose, it would be able to "air launch" rockets from 35,000 feet—and perhaps even a new space shuttle, called "Black Ice," it was developing in secret.

Together these Space Barons were behind some of the biggest brands in the world—Amazon, Microsoft, Virgin, Tesla, PayPal—that have disrupted industries ranging from retail to credit cards to air travel. And now they were betting vast swaths of their enormous fortunes that they could make space available to the masses, and push human space travel past where governments had gone.

The story of their dramatic struggle to open the frontier was an improbable one, full of risk and high adventure, a crash that cost the life of a test pilot, a rocket explosion, and suspicions of sabotage. There were lawsuits pitting an underdog upstart against the nation's military-industrial complex, a political fight that went all the way to the White House, visions to put humans on the moon and Mars, and, of course, the historic landings that heralded what Bezos was calling a new "golden age of space exploration."

At its heart, the story was fueled by a budding rivalry between the two leaders of this new space movement. The tension would play out in legal briefs and on Twitter, skirmishes over the significance of their respective landings and the thrust of their rockets, and even a dispute over the pad that would launch them. Musk, the brash hare, was blazing a trail for others to follow, while Bezos, the secretive and slow tortoise, who was content to take it step by step in a race that was only just beginning.

TIMELINE

September 2000	Jeff Bezos founds Blue Operations LLC, the precursor to Blue Origin.
March 2002	Elon Musk incorporates Space Exploration Technologies.
December 2003	First powered flight of SpaceShipOne.
December 2003	Musk shows off the Falcon 1 rocket in Washington, DC.
September 2004	Richard Branson acquires technology behind SpaceShipOne and vows to create the world's first commercial spaceline with first flights in 2007.
October 2004	SpaceShipOne wins the Ansari X Prize.
March 2005	Blue Origin flies Charon, its first test vehicle, to 316 feet.
March 2006	SpaceX attempts first launch of Falcon 1, which fails.
August 2006	NASA awards SpaceX a $278 million contract as part of the Commercial Orbital Transportation Services program.
November 2006	Blue Origin launches Goddard, a test rocket, to 285 feet.
September 2008	SpaceX's Falcon 1 successfully reaches orbit for the first time.

December 2008   NASA awards SpaceX a $1.6 billion contract to fly cargo to the International Space Station.

January 2010   President Barack Obama releases NASA budget proposal that kills the George W. Bush–era Constellation program.

April 2010   Obama gives speech at the Kennedy Space Center and visits with Musk at pad 40.

June 2010   First flight of the Falcon 9 launches successfully.

July 2011   NASA's space shuttle flies for the last time, leaving the United States with no way to launch astronauts to space.

August 2011   Blue Origin's PM-2 test rocket crashes in West Texas.

December 2011   Paul Allen announces plans to build Stratolaunch, the largest plane ever built, which would be used to "air launch" rockets.

May 2012   SpaceX's Dragon spacecraft becomes first commercial vehicle to reach the International Space Station.

March 2013   Bezos's deep-sea expedition recovers the F-1 engines from the bottom of the Atlantic Ocean.

September 2013   Tensions between SpaceX and Blue Origin heighten over Launch Complex 39A. Musk says chances of "unicorns dancing in the flame duct" are greater than Bezos building a NASA-qualified rocket that can reach orbit.

April 2014   SpaceX sues the US Air Force over right to compete for Pentagon launch contracts.

September 2014   SpaceX and Boeing win contracts to fly NASA astronauts to the International Space Station. SpaceX's contract is worth up to $2.6 billion; Boeing's, $4.2 billion.

October 2014   Virgin Galactic's SpaceShipTwo crashes in the Mojave Desert.

April 2015   Blue Origin successfully launches New Shepard to the edge of space for the first time.

June 2015	Falcon 9 explodes during launch to resupply the space station with cargo.
September 2015	Bezos announces that Blue Origin will launch its new orbital rocket from Launch Complex 36 at Cape Canaveral.
November 2015	New Shepard lands successfully for the first time.
December 2015	Falcon 9 lands successfully for the first time.
February 2016	Richard Branson unveils new SpaceShipTwo spacecraft.
September 2016	Falcon 9 explodes on the launchpad during fueling.
September 2016	Musk reveals plan to get to Mars during speech at the International Astronautical Congress.
October 2016	Blue Origin retires its first New Shepard booster after it flies and lands for the fifth time in a row.
January 2017	Blue Origin pitches NASA on a plan to fly cargo to the surface of the moon.
February 2017	Musk announces plan to fly two paying citizens around the moon.
September 2017	Musk announces plan to create a base on the moon.

PART I

IMPOSSIBLE

1

"A Silly Way to Die"

MARCH 6, 2003.

This was not how Jeff Bezos wanted to die.

He was seated in the passenger seat of a ruby-red helicopter, surrounded by an eccentric cast of characters—a cowboy, an attorney, and a pilot nicknamed "Cheater" who was best known for being forced at gunpoint to fly into the grounds of the New Mexico state penitentiary to bust out three inmates. It was a little after ten a.m. The sun had burned off the last of the morning chill as the day was heating up fast. The breeze had picked up, and with four passengers, the fully loaded helicopter struggled to lift off out of a canyon near Cathedral Mountain in the warm, thin, high-altitude air of West Texas.

Instead of going up, the helicopter began cruising along the floor of the clearing, moving faster and faster but unable to gain lift above the tree line.

"Oh, shit!" Cheater exclaimed.

In the backseat, Ty Holland, the cowboy who was serving as a guide to take Bezos around the backcountry, looked up from the topographical map he'd been studying. Bezos was sitting directly in front of him in the passenger seat, holding on; Bezos's attorney, Elizabeth Korrell, was seated next to Holland, behind the pilot. Cheater was jostling the controls, a grimace on his face, as he was "weaving and dodging between the trees," Bezos recalled.

Holland had been worried about this. The wind picks up at this time of year, swirling across the dead, desiccated desert, scattering the tumbleweed and blowing great plumes of dust. It could be especially bad up here, some 5,000 feet above the desert floor, near Cathedral Mountain, a gradual, barren incline that rises into a towering butte that from a distance looks like an elephant. But it wasn't so much the wind that was giving them difficulty. It was their weight, and the altitude, and the warm, thin air, all of which had conspired against them.

Just a few minutes earlier, Holland had urged them to get going to the next stop. But Bezos had wanted to walk around, take another look at the land, the view that carries some 80 miles to the Mexican border. The vista, miles upon miles of empty Texas desert, must have been soothing, especially for someone who led as hectic a life as Bezos. The run of the mountainside down into the desert plain, as desolate and dead brown as his hometown of Seattle was dense and lush green. The quiet of the vast expanse. Bezos had said something that morning about how he had spent summers as a kid at his grandfather's ranch in South Texas. He clearly had an appreciation for this rugged, barren country.

Holland knew little of his charge other than he was a billionaire, and that he had made his money selling books and who knows what else over the Internet on a site known as Amazon.com. He also knew that Bezos's quiet moment here at the base of Cathedral Mountain was being disturbed by a gathering breeze in the cedar trees with a sinister pitch that was making Holland nervous.

"We need to get out of here on account of the wind," he'd said. "You can't fly these helicopters up here with the wind."

Now the helicopter was in trouble. And Cheater, the pilot, was frantically trying to gain control, working the controls as if he were riding a bucking bronco at a rodeo. But there was little he could do. Best just to hold the reins and brace for impact, Holland thought. The helicopter slammed down hard and one of the landing skids caught a mound of dirt, toppling it over. The chopper's blades crashed into the ground, splintering into shards that could at any moment slice into the cabin.

Outside, the world turned upside down as the helicopter toppled into a lonely ribbon of a creek that just happened to be named "Calamity." Inside the cabin, the passengers were jostled around like pinballs, ricocheting from the force of impact, then lurching sideways as the helicopter flipped over.

The chopper's cabin lay partially submerged in the shallow creek, and water was beginning to gush inside. Somehow Holland ended up swallowing a mouthful. He did not want to survive a harrowing crash only to drown in a creek. He yanked desperately at his seatbelt. But somehow the mayhem of the crash coupled with the panic-fueled adrenaline made it impossible to undo. The seatbelt that had just saved him was now strangling him, pressing down on his chest and hips, ever tighter.

Bezos looked into the back of the helicopter to make sure Korrell was okay, but she had disappeared.

"Where's Elizabeth?" he asked, frantically.

There was no response. Then, they saw a hand rising from the water underneath Holland. During the crash, he had pinned the attorney underwater without even knowing he was on top of her. They scrambled to get her out of her seatbelt and her head above water. She gasped for breath. Her lower back was in intense pain. But she had survived. Miraculously, they all had.

They climbed out of the helicopter, one by one, gathering on the bank, taking stock. Bezos and Cheater had nicks and bruises from hitting their head against the dashboard. Korrell had broken her lower vertebrae. Holland's arms and shoulder hurt like hell. He must have torn a muscle in the crash or the scramble to get that damn seatbelt off.

Looking down at the totaled helicopter, they realized how lucky they had been. The crash had amputated the rear tail boom. The chopper lay on its side in the creek, its top rotors scalped. Fuel had spilled out everywhere, so even though Korrell had nearly drowned, the water prevented the helicopter from catching fire. Nearby, the trees were mangled as if chopped by a gardener's shears, the soil butchered—a scene altogether different from the serenity Bezos had been enjoying just a few moments before.

"It was harrowing. We were very lucky," Bezos said later. "I can't believe we all walked away from it."

FROM THE START, Holland had thought flying in a helicopter was a bad idea. Not just because he had never been in one. Or that they'd be flying into some rugged, isolated country. Holland believed that the best way to look at property in the backcountry was by horseback, his preferred means of travel. "You can get a better idea of the country by riding on horseback than flying over it in a dang helicopter," he thought.

But Bezos and his attorney "were in a big rush," Holland recalled. The trip by horse could take days. They only had a few hours.

Holland had gone on this excursion as a favor to a friend who was a real estate broker. Bezos was looking to buy a ranch, and the broker had asked Holland to show him around. No one knew the country back here as he did, and he was happy to oblige. Holland figured that Bezos, now thirty-nine years old, was looking for a place to relax on weekends, run a few cattle, and pretend to be a cowboy. Maybe relive the childhood memories of summers on his grandfather's ranch in South Texas.

Holland didn't own a computer, let alone go on the Internet. "I knew exactly nothing about him or Amazon or Internet or any of that stuff," he said.

Nearly a decade after Bezos had quit his job on Wall Street to sell books on the web, Amazon was starting to take off. In January 2002, the company had posted its first quarterly profit, $5 million. And it had continued to grow, branching out from books to music, toys, clothes, kitchen supplies, and electronics, as customers became more comfortable with using their computer to buy almost anything. In 2000, Amazon had sold 400,000 copies of the Harry Potter book released that year. Three years later, it had sold 1.4 million copies of the next installment, *Harry Potter and the Order of the Phoenix.*

Amazon was thriving at a time when many others had collapsed in a stock market swoon that claimed countless so-called dotcoms.

"We've seen the worst of the shakeout," an analyst told the *Washington Post*, after the company posted another profit in early 2003, a couple of months before the helicopter crash. "Now there are some behemoths starting to emerge."

Amazon's strategy was "get big fast," luring customers with the convenience of the Internet and the low prices that the site was becoming known for. Despite the get-rich-quick hype that had surrounded so many Internet startups, Amazon took a slow and steady approach, keeping its prices low, offering free shipping, even while critics said it would never work.

In headlines during the late 1990s, *Business Week* had derided the company as "Amazon.Toast," and *Barron's* called it "Amazon.Bomb," with an unflattering photo of Bezos, who showed it to an audience and said, "My mom hates this picture."

But by early 2003, with sales in every major segment growing by double digits, Bezos was as confident as ever in the company's approach. "It's working," he said. "It's the right investment to make, and it's in the long-term best interest of shareholders and our customers."

The iPhone was still four years away from its debut, but he was confident that the Internet was really only just getting started. In a TED Talk weeks before the West Texas helicopter crash, he compared it to the early days of the electrical industry. The web in 2003 was about where the electrical industry was in 1908, he argued, when the electric socket hadn't yet been invented and appliances had to be plugged into light sockets.

"If you really do believe it's the very, very beginning," he said, "then you're incredibly optimistic. And I do think that's where we are."

With Amazon's success, Bezos's wealth was growing rapidly. *Fortune* magazine reported in 2003 that with Amazon's stock price tripling, his net worth grew by $3 billion to a total of $5.1 billion. He vaulted to number 32 on the list of wealthiest Americans, ahead of New York media titan Michael Bloomberg and the Koch brothers, who ran a vast manufacturing and investment empire.

March 2003 was, then, a good time to go looking for real estate. To allow himself a measure of freedom to indulge his true passion, even if he rarely spoke of it.

BEZOS DIDN'T SAY anything about why he wanted to buy land in this hideaway corner of Texas, full of rattlesnakes, mule deer, bighorn sheep, and not much else. Holland, a stolid, soft-spoken rancher who spent as much time around cattle as people, didn't ask. Bezos struck him as a "different breed of cat," one he couldn't relate to.

Suspicious as Holland was about the helicopter, he was also wary of its pilot, Charles Bella, who was something of a legend in his hometown of El Paso. He had a handlebar mustache and a penchant for fistfights and profanity. His nickname, "Cheater," came from his days of racing cars: sore losers had accused Bella of cheating and the nickname had stuck. "It turned out to be a compliment," he told a magazine in 2009. He'd been hired by Hollywood to fly in several movies, including *Rambo III* and a Chuck Norris flick called *Lone Wolf McQuade*. In addition to his work as a helicopter pilot, he kept a gaggle of exotic animals—including a bear, timber wolves, mountain lions, and an alligator—at his home. The local game warden would call on him from time to time, once telling the *El Paso Times* that "Cheater has a way with animals. He can go into a cage with injured mountain lions, and they turn into pussycats."

What he was best known for, though, was the prison break. In 1988, he had flown his Gazelle helicopter—incidentally, the same one he flew in *Rambo III*—into the prison, freeing three inmates. After a two-hour chase, he was arrested and charged with conspiracy. But F. Lee Bailey, the famed criminal lawyer who would go on to be part of O. J. Simpson's defense, represented him and mounted a vigorous defense that led to his acquittal.

The morning of the prison break, Cheater claimed, a woman had hired him to look at some real estate, just as Bezos later would do. She was dressed in bright red pants and a floral print shirt. That morning, she had taken several guns from a roommate, leaving her a note that read, "Katie, I'm taking your guns because I need them more than you do."

Soon after they took off, she took out one of the guns, a .357 magnum, pointed it at Cheater's head, and demanded he fly into the prison to free her boyfriend, a convicted murderer serving a life-plus-sixty-years sentence, and two of his friends.

She was obese, Cheater recalled, about 250 pounds. "I'm thinking I'm in deep shit because if this ol' gal thinks that guy loves her, she won't stop at nothing, because she'll never find another guy," he told *Texas Monthly* years later.

He said he tried grabbing at the gun, but his hand was sore from a fist-fight days before and he couldn't wrest it free. They landed on the prison ball field near first base, where the three prisoners were waiting. They scrambled aboard, one hanging on to the helicopter skid, while guards fired from the prison tower. Cheater wasn't sure what to do.

"Her boyfriend is slapping me on the head with a gun saying he's gonna blow my head off if I don't get going," he recalled. "The engine is already up to the max. I'm pulling it all the way through the temperature range. It should have exploded. And finally they shove this one guy off the skids and one of the guys in the helicopter jumps off and runs alongside it and he climbs back in when we start to take off."

The chopper barely cleared the fence, and they were off. Soon they had bigger problems. The feds were in pursuit, chasing them down in a Black Hawk helicopter for nearly two hours until it was clear there was no escape. Cheater finally landed at the Albuquerque airport.

Now, with his helicopter lying broken and submerged, he was in another big mess. He had just crashed in the middle of nowhere with one of the world's richest people on board.

Bezos's entourage was wet and stranded, miles from civilization, with no service to call for help on their cell phones. Still, it could have been much worse. They were alive, and as they stood on the banks of Calamity Creek, a modicum of relief settled in. Bezos looked at Holland and smiled.

"Maybe you were right," he said to the cowboy. "Maybe horseback was the way to travel into the backcountry after all." A weird, full-throttle laugh boomed through the canyon from the man who had cheated death.

"He let out that goofy laugh," Holland said. "He thought it was funny as hell. I didn't think it was funny."

Cheater set off the helicopter's transponder, hoping rescue crews would pick up the signal. And Holland took off on foot to look for help at a house a few miles away.

They didn't have to wait long. Soon, helicopters from US Border Patrol appeared overheard, and then the Brewster County sheriff's deputies showed up with a backcountry rescue group.

Sheriff Ronny Dodson surveyed the scene. The chopper was a red carcass, slumped in the creek. The scarred ground that had been plowed by the helicopter's rotors. Then this curious quartet. As a lawman, he was familiar with the infamous Cheater Bella. And he recognized Holland as one of the local ranchers. But the short, quirky-looking fellow seemed out of place. Friendly as Bezos was, Dodson couldn't place him.

It wasn't until paramedics showed up in a pickup truck and one was amazed to see the 1999 *Time* magazine Person of the Year at the crash site that Dodson was clued in.

"Don't you know who that is?" said one paramedic to the clearly oblivious sheriff. "It's the founder of Amazon.com."

Amazon? Yes, Dodson had heard of it, though he wasn't a customer.

"I didn't do much business with Amazon," he said, years later. "I thought Amazon was just books, and I read about none. So, why would I look?"

The paramedics took Bezos and Korrell to the hospital: he was treated for minor cuts; she, for her broken vertebrae; both were released. Holland's arms were still hurting from trying to rip off the seatbelt. "I lived in pain my whole life, and knew something was wrong," he said. The doctor said he'd need to see a specialist. But he'd had enough for the day.

"I got my shirt on and left," he said. "And went to the bar."

As word of the crash spread, Amazon downplayed it, declining to comment except to say that Bezos "is fine. We're business as usual." Years later, Bezos would admit that the crash was far more severe, but still made light of his brush with death.

"People say that your life flashes before your eyes," he told an interviewer at *Fast Company* magazine in 2004. "This particular accident happened slowly enough that we had a few seconds to contemplate it."

He let out his trademark, maniacal laugh. "I have to say, nothing extremely profound flashed through my head in those few seconds. My main thought was, 'This is such a silly way to die.'

"It wasn't life-changing in any major way," he continued. "I've learned a fairly tactical lesson from it, I'm afraid. The biggest takeaway is: Avoid helicopters whenever possible. They're not as reliable as fixed-wing aircraft."

It wasn't long after the crash that Ronald Stasny's phone began to ring. The attorney on the phone was polite, unfailingly so, but also incredibly persistent. Every month or so, Elizabeth Korrell said she was calling on behalf of a mysterious client, one she refused to name, and every time, Stasny's response was the same. No, he was not interested in selling his ranch.

It had a great view from the kitchen window of Guadalupe Peak, the highest point in Texas, and it was surrounded by the ethereal mountain ranges of West Texas, the Sierra Diablos, the Baylors, the Apaches, and the Delawares surrounding the ranch like sentries. Stretching out 32,000 acres, it was home to quail, dove, and mountain lions. The mule deer were big and plentiful, with massive antlers that branched out like leafless trees, and fed on the protein-rich vegetation. Stasny's grandchildren were gaining a lifetime's worth of memories discovering the far-flung secrets of this swath of Texas plain, from the old silver and gold mine shafts to the Indian artifacts.

No, he wasn't going to sell. Especially not to some Seattle lawyer with a secretive buyer. This was Stasny's retreat, a place to get away from the city life of San Antonio, where he was a lawyer himself. This was where he and his wife planned to retire.

The old adobe house on the property dated to the 1920s, and the lineage of the land was tied to one of the great ranches in all of Texas, the Figure 2 Ranch, where the Texas Rangers had fought one of their last battles against the Apaches in 1881. The land had passed through the family of James Madison Daugherty, one of the founding fathers of Texas cattle ranchers, and James Marion "Silver Dollar Jim" West Jr., a Houston oil scion, who had a habit of flipping silver dollar coins to people on the street.

Stasny had invested a great deal in the property, adding heat and air-conditioning to the main house. When a hailstorm left pockmarks in the roof, and he wanted it replaced, the roofer urged him not to, saying, "They don't make sheet metal roofs like this anymore." He put in an irrigation system and cleared some of the paths to the backcountry that had been accessible only by horseback. The hunters were grateful that he allowed them on the property, staying ten deep in the bunkhouse behind the barn, before stalking those glorious mule deer.

But Korrell, the Seattle attorney, wouldn't give up. Her anonymous client was very interested in the property. And as Stasny later would learn, he wasn't the only landowner they were courting. Korrell was reaching out to several of his neighbors, offering to buy them out as well. Here was someone who could clearly afford to buy it all up—and was, judging by the frequency of the attorney calls, eager to do so. One by one, his neighbors sold.

Finally, Stasny did, too. He accepted the deal in early 2004, after talking about it with his family over the Christmas holiday. The offer was just too good, more than enough to buy another ranch to retire on. Although he wouldn't say what he sold for—he signed a confidentiality agreement—it was reportedly $7.5 million.

The mysterious buyer was amassing an impressive collection of ranches, cloaking his identity by buying the land under corporate entities with curious names: Jolliet Holdings and Cabot Enterprises, the James Cook and William Clark Limited Partnerships, and Coronado Ventures. All were named for explorers who had opened up frontiers, from the American West, to New Zealand and the Great Barrier Reef, to Mexico and Canada. All linked to a little-known corporation, doing business out of Seattle Post Office Box 94314, with an out-of-this-world-sounding name: Zefram LLC.

Right there was a clue to the buyer's identity and his intentions. Zefram Cochrane was a character in *Star Trek* who created the first spaceship capable of traveling at warp speeds, or faster than the speed of light. He was an explorer of another sort, a fictional one from the future, who said the warp speed engine "will let us go boldly where no man has gone before."

RUMOR WAS STARTING to spread around the nearby town of Van Horn that someone was out there buying up property in Culberson and Hudspeth Counties. Larry Simpson had a pretty good idea who.

Simpson was the owner and editor of the *Van Horn Advocate*, the weekly newspaper that he ran out of the back of his office supply store with a handful of reporters for the town of 2,100. Simpson also worked part-time at the county airport. Word there was Bezos had been flying in on his private jet with a real estate agent in tow. The helicopter crash had gotten people wondering as well. But what the billionaire wanted the land for was anyone's guess.

Simpson didn't give it much thought. "I have not been real pushy, like maybe a big-city newspaper guy would be," he told the *Seattle Times*. But then one Monday in January 2005, Bezos himself stopped by the office, saying, "We want to give you a news release." Casual, relaxed in jeans and boots. No entourage, just Bezos and a soft-spoken gentleman who let his boss do the talking.

Bezos had a scoop for the *Van Horn Advocate* and its circulation of one thousand: he was buying all the land for his space company, a little-known venture called Blue Origin, based outside Seattle. Since its founding in 2000, Bezos had kept the company extremely secret, telling virtually no one what its plans were. The company wasn't listed in the phone book. Its employees told neighbors they were working on scientific research. And one industry official told the *Economist* magazine that "everyone I know who knows anything about it isn't allowed to talk about it. And please, please don't quote me on that."

After reports of the helicopter crash, Brad Stone, then a young reporter at *Newsweek*, became interested in what Bezos was up to. He found filings for Blue Operations LLC in Washington state records and visited a warehouse in Seattle's industrial area late one evening, as he wrote in *The Everything Store*, his book about Amazon. After sitting outside the facility for an hour, he scooped up a bunch of documents from a trash bin, which included a coffee-stained mission statement to create, as Stone wrote, "an enduring presence in space."

Stone reached out to Bezos for comment, but Bezos declined to elaborate on Blue Origin's goals for Stone's *Newsweek* story, titled "Bezos in Space."

"It's way too premature for Blue to say anything or comment on anything because we haven't done anything worthy of comment," Bezos wrote in an e-mail. He did, however, seek to dispel one falsehood—that somehow he was doing this out of a frustration with NASA, which many had criticized for going backward after the Apollo moon missions.

"NASA is a national treasure, and it's total bull that anyone should be frustrated by NASA," he wrote. "The only reason I'm interested in space is because they inspired me when I was five years old. How many government agencies can you think of that inspire five year olds?" Bezos was five when Neil Armstrong and Buzz Aldrin landed on the moon in 1969.

FOR A WHILE, the company's only employee was Bezos's friend Neal Stephenson, the science fiction author. They had met in the mid-1990s at a dinner party, where they had started talking about rockets. While the pair may have "bored everyone else at the table," Stephenson said, they hit it off. "It was super obvious he knew a lot."

As their friendship grew, they spent time launching model rockets in Seattle's Magnuson Park overlooking Lake Washington. Once, one of the rocket's parachutes got tangled in a tree as it fell back down, "and before I knew it, he had scampered into the tree and had literally gone out on a limb," Stephenson recalled. It wasn't very strong, so Bezos tried "shaking the limb with his body weight." His wife, MacKenzie, was pleading with him to come down, when a dog walker came by and knocked it out with a stick.

In 1999, Bezos and Stephenson went to see a matinee of *October Sky*, the film about writer and NASA engineer Homer Hickam. Bezos smuggled peanut butter sandwiches into the theater by hiding them in his jacket, and afterward, in a coffee shop, he said that he had always wanted to start a space company.

"And I said, 'Well, why don't you start it today?'" Stephenson recalled.

Why *not* start it today? What was he waiting for? He'd been fascinated with space his whole life and now he was finally in a position to do something about it.

"Things started immediately after that," Stephenson said.

Stephenson would be the first hire, and he introduced some friends to Bezos, who hired them as well. They were "the sorts of people who would be good at rapidly getting their heads around weird applied physicist ideas and evaluating them," Stephenson recalled. Their titles were simple and egalitarian: "member, technical staff." Stephenson held a variety of roles "as prosaic as punching down Ethernet cable, operating a plate grinder, and passivating rocket parts (which means making sure they don't have any residue that would react with hydrogen peroxide)," he wrote on his website.

Bezos acquired the building at 13 South Nevada Street in Seattle's industrial section, where they formed a sort of think tank dedicated to space. Stephenson worked part-time, writing in the mornings and then working at the facility during the afternoons. Bezos kept tabs on the group, consulting with them frequently and coming in for meetings one Saturday a month. Their first goal was to explore other ways of getting to space besides using chemical rockets, a technology that hadn't improved much in the four previous decades. For the first three years "we exhaustively looked at every known alternative to chemical rockets and we even invented some previously unthought-of alternatives to chemical rockets," Bezos said.

To develop a whole new technology, they'd consider any possibility, no matter how crazy it sounded. "When you're brainstorming you have to accept wild ideas," Bezos said.

The wildest idea was probably the bullwhip. An ancient technology but an amazing one: as the loop of the whip unfurled into the distance, it would generate so much velocity that it could actually break the sound barrier.

"How are you able with just your arm to get something moving above the speed of sound, well, it's conservation of momentum," Bezos said later. Momentum is mass times velocity. As the whip gradually got thinner from the handle to the tip, its mass decreased, meaning that to conserve its momentum, the velocity would have to increase. So much so that the whip's trademark *crack* was actually a sonic boom.

"And so we said, 'Why don't we just make a giant one of those?'" Bezos continued. "It would be a gigantic bull whip where you would put the thing you wanted to fling into orbit at the tip of the whip. So, picture a space capsule or a payload or whatever you wanted to put on the tip of the whip."

There would still need to be a rocket engine on whatever was being flung into space, to give it the velocity needed to reach orbit. But the speed of the whip would allow the spacecraft to carry more mass, and farther. The whip would obviously have to be gigantic—"freight trains would have to whip this thing," he explained. "It had all sorts of practical problems. It's the kind of thing you could dispose of as a credible idea in a few hours of analysis. So as far as we know, nobody had ever considered that."

So, instead, they then studied what Bezos called "much more reasonable things with much more seriousness."

Such as lasers.

There would be a field of lasers on the ground that would keep a constant beam on the rocket as it raced across the sky, heating the liquid hydrogen propellant, which has a very high specific impulse, or efficiency. The company was serious enough about it that it hired a consultant, Jordin Kare, who prepared a study. The problem was that to generate the necessary energy, you'd need a massive field of lasers, "so, this is a little impractical from a cost point of view," Bezos said.

But, in theory, it would work. And maybe if laser technology improves there could be laser-beamed rockets shooting off into space. But not now.

If lasers didn't work, maybe giant space cannons would—"ballistic solutions where you have big gun barrels of various kinds," Bezos said, blasting objects into space as if in a Jules Verne novel. This, obviously, would not be good "for humans because the G [gravitational] forces are too big," he said. But it could work for "getting cargo up to space."

So, they looked at railgun technology, which was being researched by the Pentagon for weapons that could fire projectiles at Mach 7, or seven times the speed of sound. (For comparison, a Hellfire missile travels at Mach 1.) Instead of using gunpowder, railguns use electromagnetic pulses; the projectiles hit with such force that they don't need to be armed with explosives.

After three years of research, Bezos and his small team ultimately decided that "chemical rockets actually are the best solution," he said. "They're not just a good solution, they're actually an awesome solution for launch."

But there was a caveat: they had to be reusable. Rockets had been to this point largely expendable. The first stages would boost their payloads into space, separate, and fall back to Earth where they would crash into the ocean, never to be used again. Each launch would require an entirely new rocket, and rocket engines, crafted carefully for a single launch. They were like honeybees sacrificing their lives to use their stinger a single glorious time. All in to the death. But what if rockets didn't need to be that way? What if they could fly again and again, like airplanes, instead of ditching into the depth of the ocean, where they would corrode and waste away?

This, Bezos thought, was the solution they had been searching for.

"When the company made a decision to stick with a more tried-and-true approach," Stephenson wrote, "I found other ways to make myself semi-useful, largely in the realm of trajectory analysis, until I decided to make an amicable withdrawal in late 2006."

In his novel *Seveneves*, he even wrote Blue Origin and the bullwhip concept into the narrative, and he dedicated the book "To Jeff," among others.

FOR ITS FIRST few years, Bezos had said virtually nothing publicly about Blue. Until this sudden trip to the *Van Horn Advocate*. Sitting across from Simpson, the editor, Bezos laid out the company's plans. The news broke when the *Advocate* published its story on January 13, 2005, under the headline "Blue Origin Picks Culberson County for Space Site."

"Blue Origin, the Seattle-based business venture today announced its plans to build and operate a privately funded aerospace testing and operations center on the Corn Ranch, north of Van Horn," Simpson's report began. The front-page story ran next to an ad for the 56th Annual County Livestock Show.

It quoted Bezos as giving a canned corporate line that Texas "has been a long-standing leader in the aerospace industry, and we are very excited about the possibility of locating here."

But Simpson's article also contained some insight into Blue's goals, especially its desire to develop "vehicles and technologies that will help enable an enduring human presence in space." Blue intended to build rockets that would take three passengers or more on suborbital flights to the edge of space. The story didn't use the phrase *space tourism*, but that's what Bezos was describing, and what he had in mind.

Blue's rockets would also do something different—something no rocket had ever done before. After blasting off from the launchpad, they would, as Simpson wrote, "land vertically."

Reusable rockets were a dream that had eluded the space community for years. The government had tried this, and failed. Its Delta Clipper Experimental, or DC-X, rocket had flown several times in the 1990s, flying to several thousand feet, before touching down softly. But never had they flown to space and then landed.

Stephenson was convinced it was achievable. "How is the situation in the year 2000 different from 1960? What has changed?" he said. "The engines can be somewhat better, but they're still chemical rocket engines. What's different is computer sensors, cameras, software. Being able to land vertically is the kind of problem that can be addressed by those technologies that existed in 2000 that didn't exist in 1960. So the story added up on that level."

In the interview with the *Van Horn Advocate*, Bezos made it clear that "it may take several years for this project to get in high gear." But his team was patient, and had hired some of the best engineers in the country, some of whom had been involved in the DC-X program and other private space ventures that had failed. He also stressed that the money for the venture was coming out of his pocket—not the taxpayers'.

"This is a privately funded, non-governmental enterprise," he told Simpson.

By now he was worth $1.7 billion, climbing higher on the *Forbes* list of the country's richest people. Earlier, scientists at the California Institute of Technology had invited Bezos and Stephenson to lunch, as part of a fund-raising drive for a new telescope. But they were not successful in getting Bezos to open his wallet.

"It became obvious that Blue Origin was where Bezos was putting his money," said Richard Ellis, a Caltech scientist at the time, who had sat next to Blue Origin employees at the lunch. "Those guys wanted to sell the concept of human space travel. They said, 'If we think outside the box, there's going to be a revolution.'"

The revolution would begin in West Texas, where Bezos would eventually acquire 331,859 acres, according to land records. That's nearly half the landmass of the state of Rhode Island. "When you are building rockets and launching rockets, it's nice to have a bit of a buffer," he once told the television host Charlie Rose.

He had that and more now. A ranch for his family, not unlike his grandfather's in South Texas where he'd spent summers and learned the value of self-reliance. A place to launch and land rockets, large enough to hold even the biggest of dreams. A place to stretch out toward the stars.

2

The Gamble

T HE DEEP-POCKETED SUCKERS who show up on the Las Vegas strip with more money than skill and view poker as mere entertainment rather than big business, are known to the serious players as whales. Fat, fleshy targets. A chance to score big. So when one wades unsuspecting into the shark-filled strip, word spreads.

When Andy Beal arrived at the Bellagio in early 2001, Vegas had never seen a whale quite like him. Tall and broad with bright eyes and a gentle smile, he sat down at the high-stakes poker table with a tall stack of chips, eager to play. The blinds were $80 and $160, meaning the minimum bet to stay in the hand was $160. After a few raises, the pot would quickly grow to well over a thousand dollars.

Beal played a few hands and grew bored. He had flown in from Dallas to gamble big, and was itching for some action. "Doesn't anybody play higher around here?" he asked.

With that, the whale made himself known. Fresh meat at the Bellagio. The next day, the professional players were waiting for him to show up, including some who had competed in the World Series of Poker. They started with a $2,000 minimum bet. But after about a half-hour, Beal was getting itchy.

"Can we play bigger?" he asked.

So, they raised the stakes to $6,000. That was fine for a while, then Beal asked to go higher, this time to $8,000.

At the table, Jennifer Harman was starting to sweat. She was a pro who would go on to win two World Series of Poker bracelets, but she had never played such high stakes. This whale was aggressive, his style unrestrained, and he rarely folded, driving up the pot every time. With so much money at stake, the pressure got to even the veteran players who couldn't keep pace.

"This was not a comfort zone for anyone in the game," Harman recalled later. "Because this is the highest that anyone has ever played."

This whale was smart and cunning and relentless. He pressed the one advantage he had, his immense wealth, until the exorbitant piles on the table stretched them thin. They were beginning to think maybe he was not a whale after all.

In fact, he was a billionaire real estate investor whose Texas bank was one of the most lucrative in the country. Beal had dropped out of college but was a genius who dabbled in number theory as an amateur mathematician. After studying Fermat's Last Theorem, a problem that has baffled mathematicians since 1637, Beal came up with a problem of his own. Given its complexity, it eventually got the attention of scholars in the field, who dubbed it the Beal Conjecture and were stunned that someone with no formal training could devise such a problem. (The conjecture was to prove that if $A^x + B^y = C^z$, where A, B, C, x, y, and z are positive integers and x, y, and z are all greater than 2, then A, B, and C must have a common prime factor.)

"It is remarkable that occasionally someone working in isolation and with no connections to the mathematical world formulates a problem so close to current research activity," wrote R. Daniel Mauldin, a professor of mathematics at North Texas University.

In 1997, the American Mathematical Society sponsored a contest for whoever could solve the problem, and Beal offered a $5,000 prize. The pot grew to $10,000, then $15,000, as Beal pledged to increase it by $5,000 every year it went unsolved up to $50,000. In 2013, the problem was still unsolved and Beal upped the jackpot to $1 million.

At the Bellagio, the pros, who included some of the biggest names in poker, such as Doyle Brunson and Ted Forrest, had decided that to afford to compete against him, they'd have to pool their resources and take turns playing the Dallas banker one-on-one. Together they formed what they called the Corporation. But during their first match, Beal wiped them out, winning some $5 million.

"We're broke," Brunson said. "Congratulations. Go back to Texas. Come back another time, and maybe we'll have some money to play you."

But Beal didn't want to go home. He had come to Vegas to play.

After losing $5 million, most people would have thought better about taking the bait. But the Corporation was made up of people who gambled for a living, and they couldn't resist the possible upside. They reached out to their contacts, and soon were able to borrow tens of thousands of dollars each from people who knew they were good for it. Within the hour, $1 million was delivered to the Bellagio high-stakes poker room. The Corporation was back in business—now more familiar with Beal's particular brand of play, his style and cadence. After an epic all-nighter, the Corporation won back its $5 million.

Over the years, Beal would return to Las Vegas again and again, taking on the Corporation in some of the most legendary games in Vegas history. During one trip, in 2004, Beal won $11.7 million. On another run, he took $13.6 million from the pros. But later lost $16.6 million in a game.

What the Corporation didn't know was that the games were a form of catharsis for Beal, a balm against a much bigger loss, where the stakes had been far higher and more significant. Beal had recently started a space company—without any experience in aerospace or engineering or rocket science. Skeptics said he was crazy. That starting a space company from scratch couldn't be done. That he was going to waste a fortune on what was little more than a midlife crisis.

Only nations could fund space programs. The prospect of a commercial outfit's succeeding was about as remote as Mars. Beal didn't care. He was convinced that he could start a space company on his own without federal funding. No matter that it hadn't been done before. Or that space had been the exclusive domain of governments, not private companies founded by amateurs. As in Vegas, he knew the longest odds led to the greatest payoffs.

At the poker table, anyone could catch a hot streak of cards and get lucky. The space business required step-by-step precision and rigor. It was unforgiving, his industry friends told him. They reminded him of the running joke in the space community, which had attracted all sorts of fantasists whose starry dreams ended up in ashes:

What's the quickest way to become a millionaire in space?

Start out as a billionaire.

SOON AFTER FOUNDING Beal Aerospace in 1997, Beal took over a former military test range in McGregor, Texas, that was once called the Bluebonnet Ordnance Plant, after the Texas state flower. The sprawling site was part of the World War II effort to build munitions, producing bombs that ranged in size from 100 to 2,000 pounds. While McGregor was once a sleepy town that sat at the junction of the Santa Fe and Cotton Belt railroads, it burgeoned with Bluebonnet's wartime expansion.

"Bluebonnet was like a city unto itself, complete with living quarters, security and fire protection services, stores, merchants, entertainment venues, bus service and a regular plant newspaper," according to a history of the site.

After the war, the US Air Force took over, expanding the site and using it to test solid rocket propellants and rockets that would be attached to the outside of aircraft in an effort to give them an extra boost. Over the years, the military would also test motors for missiles, and by the time the plant shut down, it had produced more than 300,000 rocket motors for the Pentagon and NASA.

It sat vacant until Beal Aerospace came looking for a test site. Despite the naysayers, Beal had grand plans to build the most powerful rocket engines since the ones that had powered the Apollo-era astronauts to the moon. Ever the opportunist, he saw space as a business that had been dominated by NASA and the US military, one that had grown sclerotic under the government's stifling monopoly. That meant, as he would say later, "it was a good opportunity. There was a lot of room for improvement."

In founding Beal Aerospace, his plan was simple: "Come up with a rocket that doesn't cost $200 million to launch." He could build rockets that would cost far less and undercut the market, disrupting an industry ripe to be upended.

This was the approach he had adopted since he was an eleven-year-old kid in the 1960s, buying broken televisions from the Salvation Army for $1 with his Uncle Denny, who taught him how to fix them and then sell them for $40. When he was nineteen, he bought a house in his hometown of Lansing, Michigan, for $6,500 and rented it for $119 a month. After high school, he took a year off to pursue real estate investments.

He enrolled in Michigan State University, mostly to appease his mother, who wanted to see his formidable intellect sculpted in a formal setting. But Beal was anxious to get out into the real world, and his heart was in his growing business. Soon he amassed fifteen properties that he renovated and rented out.

He ended up dropping classes, or taking incompletes for his many absences—and then leaving school entirely. It proved to be a lucrative move. When he was in his early twenties, he bought a federally owned apartment complex in Waco, Texas, for $217,500 at a government auction, even though he hadn't seen the complex—and had never been to Texas. Three years later, in 1979, he sold it and pocketed more than $1 million. Within a few years, he bought and flipped property after property, making himself a millionaire several times over.

When the savings-and-loan crisis hit in the late 1980s, he found a way to profit from that as well. He started a bank with $3 million of his own money, and started buying up loans at discounted prices. "If everybody else is going broke, that simply means your competition is going away," he told a Dallas magazine in 2000.

Beal Bank became one of the most profitable in the country, with more than $1 billion in assets by the mid-1990s. Now he was wealthy enough to pursue his interests beyond Earth.

BEAL LEARNED ABOUT aerospace the same way he had learned much throughout his life: he taught himself. He read every book on rocket science, engine mechanics, propulsion. He talked to engineers and scientists and founded Beal Aerospace with the goal of dramatically reducing the cost of space travel.

He took on entrenched contractors, such as Lockheed Martin and Boeing, which had built their businesses by managing the byzantine federal government contracting bureaucracy as much as by creating innovative technologies. And he saw a boom in satellite technology coming, a new market that would need companies that could launch objects to orbit faster and cheaper.

But he was also drawn to the business for another reason, one that would have raised eyebrows in his conservative Texas banking and real estate circles. He feared for the future of humanity. At some point Earth could be hit by an asteroid, wiping out the human race the way the dinosaurs had gone extinct. If humanity was to survive, he thought, it would need to find a way to live on other planets in the solar system.

"I don't lose a lot of sleep over that," he said of an asteroid collision, "because it could be a billion years or hundreds of millions of years or tens of millions of years away. But the fact is, it could be 20 years away. And to the extent that our efforts speed up the colonizing of other planets—you just never know all the implications of efforts like this. . . . So all the knowledge, all the answers to questions that we don't even know to ask, all that would be furthered by what we were doing."

Beal hired some of the best engineers in the country, poaching them from Lockheed, Boeing, and Orbital Sciences. They began to make a massive heavy-lift rocket, the BA-2, which would have been the largest rocket engine since the F-1 engine that powered NASA's Saturn V rockets during the Apollo era. The three-stage rocket would stand 236 feet tall and have the strength to carry nearly 20 tons to low Earth orbit.

In early 2000, the company christened the McGregor facility with a successful test of its second-stage engine, the largest liquid-fueled engine built since the Apollo program. It roared to life before a crowd of a couple hundred, sending a fiery blast that consumed 63,000 pounds of propellant in just twenty-one seconds.

But as his company grew, Beal grew concerned about its prospects. According to reports at the time, he had spent about $200 million of his own money on the venture without taking a cent of taxpayer money from NASA or the military. What worried him most was not the engineering challenge of building such a massive rocket, or the perils of space, but the federal government's lock on the industry.

NASA and the Pentagon had programs of record, working with such stalwarts as Lockheed Martin and Boeing, and weren't particularly interested in giving contracts to unproven upstarts like Beal Aerospace. Beal didn't think it was fair to have to compete against companies that had been subsidized by the US government.

He took his concerns to Washington, where at a 1999 Senate hearing, he testified that "while we are confident of our ability to compete on a level playing field, one of our biggest risks is that well-intentioned government actions might improperly tilt the playing field by rewarding or penalizing various competitors, essentially predetermining the winners and losers."

The federal government should be in the business of buying services from companies but not helping them build their rockets, he said. Billions of dollars spent on big government programs might be good pork for particular districts where the rockets are manufactured, but they run against free-market forces and would not yield anything, he warned.

"PLEASE, PLEASE do not give companies billions of our dollars to play around with experimental programs," he wrote in his testimony to the Senate committee. "You will create jobs by spending public money, but you absolutely will not produce low-cost commercial access to space."

Then in 2000, months after the company had successfully test fired one of its engines—a fiery projection of force designed to show NASA and the aerospace industry that the company was legitimate—the space agency went and did what Beal had feared it would. It announced it would pursue a multibillion-dollar program, known as the Space Launch Initiative, for the development of space vehicles designed to replace the shuttle and be reusable.

Beal saw it as a fatal blow. There was just no way to compete with companies that were subsidized by the government. He had hoped that the space industry would become a truly commercial one, where the government was yet another customer—not the only one. But that day was still years away.

In October 2000, Beal announced that the company would be "ceasing all business operations," effective immediately. He championed his company's technical prowess, the "significant advances in low cost hydrogen peroxide propulsion systems." Company officials were "confident of our ability to ultimately succeed in the development of our BA-2C rocket launch system," which he noted was the "largest privately funded program ever in existence to build a large capacity space launch system."

But then the press release read more like a prophecy than a corporate announcement: "There will never be a private launch industry as long as NASA and the U.S. government choose and subsidize launch systems."

The government, he said, needed to get out of the way, and let the free market take off. Only then would NASA's monopoly on space end, touching off a new space economy.

"We wonder where the computer industry would be today if the U.S. government had selected and subsidized one or two personal computer systems when Microsoft, Inc. or Compaq, Inc. were in their infancy," he wrote.

Perhaps one day NASA would be ready to open its doors to the commercial sector. Perhaps his efforts had put a dent in the ceiling, paving the way for the next starry-eyed venture. But as Beal's failed effort showed, mastering the art of rocket science would not be enough. The next person who desired to start a space company would have to wage war—in Washington, in court, and in the court of public opinion—against the entrenched interests Beal could not overcome. It seemed that starting a successful space company was well beyond even the most hopeful dreams. It was a delusion.

Beal had decided he had two choices: join the establishment and "evolve into a government contractor role like Boeing and Lockheed" that built the systems NASA and the Pentagon asked for, or shut down entirely.

Beal knew a bad hand when he saw one. He decided to fold.

WHEN BEAL AEROSPACE went away, McGregor had lost a valuable tenant—and contributor to its coffers. Now, the city suddenly had a few hundred industrial acres on its hands for lease, and no prospects. Who in their right mind would want to take over a site that was good for testing rockets and little else? Who else would even try to start a rocket company? Beal, the billionaire math genius, had tried and failed, and his was now a cautionary tale that proved the skeptics right.

The McGregor parcel had been a testament to his bold gamble, a symbol of a new, exciting vision of private space. But more than a year later, as the acreage sat vacant, snakes and scorpions moved in. The brush spread. The testing stands started to rust. The site represented abandoned dreams, a wasteland that seemed destined to further degrade under the searing Texas sun.

Then, in 2002, the McGregor city manager got a curious call from a man named Jim Cantrell, who had been scouting sites for his boss. He had looked in the Mojave Desert and in Utah, remote locations where they could operate without worrying about environmental concerns. But none of them had worked out. Then, Cantrell remembered seeing a picture in *SpaceNews* when it did a story on one of Beal's engine test firings. He looked up the story, which identified the McGregor test site.

The city manager was affable and accommodating, with a thick Texas drawl.

"How can I help you?" he asked Cantrell.

Cantrell said he was interested in the site where Beal had tested his rockets and was hoping to learn whom he might talk to about taking a look at it.

"Well, you're talking to the man who owns it," the city manager said.

Cantrell told him that he worked for a man named Elon Musk, who had made a lot of money on the Internet and had started a company called Space Exploration Technologies. Never heard of him, the city manager said. Still, anyone interested in the property was welcome to visit anytime.

They flew in on Musk's private jet, a Falcon Dassault 900. Musk looked around and made up his mind quickly. "This is perfect," Cantrell recalled him saying. Musk signed the lease, and starting in 2003, he had 197 acres, a test stand, and five buildings for $45,000 a year.

Musk was, in many ways, like a younger version of Beal. Instead of repairing and reselling televisions, Musk's childhood entrepreneurial streak led him to try to open a video arcade when he was sixteen with his brother Kimbal in their hometown of Pretoria, South Africa. But they were stopped by the city because of a zoning issue. "Our parents had no idea," Kimbal said. "They flipped out when they found out, especially my father."

Elon had a rough childhood, and a strained relationship with his father. It was clear that from a young age he was exceptional, and his mother sent him to school early, making him the youngest, and smallest, in his class, a prime target for bullies. "It's pretty rough in South Africa," Kimbal told *Esquire*. "It's a rough culture. Imagine rough—well, it's rougher than that. Kids gave Elon a very hard time, and it had a huge impact on his life."

Elon Musk fled South Africa after high school when he was seventeen. First, he went to live with relatives in Canada, where he enrolled in Queen's University. Then, he transferred to the University of Pennsylvania, graduating with degrees in physics and economics. He had been accepted at Stanford, where he planned to study the technology behind ultracapacitors, hoping to create a better battery that could be used in electric cars.

It was 1995, the dawn of the Internet era. "I thought the Internet would be something that would fundamentally change the nature of humanity," he said during a speech in 2012. "It was like humanity gaining a nervous system." So, he told his professor he was taking a deferment to see whether he could start an Internet company, and the teacher said, "Well, I don't think you'll be coming back." That was the last time they ever talked.

He started a company called Zip2 that would help print newspapers get their content online, and it immediately had customers lining up, from the *New York Times* to Hearst. Musk sold the company to Compaq in 1999 for about $300 million. His next venture was called X.com, an online bank that merged with PayPal. The online financial payment system grew fast, gaining a million customers within two years, "and we didn't spend any money on advertising." In July 2002, eBay bought PayPal for $1.5 billion, netting Musk $180 million. He was thirty-one years old.

Even before the sale, Musk was thinking about what he wanted to do next, what benefit he could make to the future of humanity. Beal had said he wanted his efforts to contribute to people's ability to stretch out into the stars—and to stay. Musk, too, thought something needed to be done.

What if the sun burned out? Or an asteroid hit Earth?

THERE WERE LOTS of big rocks out there, such as the one NASA spotted in the mid-2000s that was big enough to fill a college football stadium. At first, it looked like a fuzzy smudge in the great distance, but NASA's astronomers didn't like what they saw. The asteroid was on an orbit to come dangerously close to Earth, so close that it would barrel by beneath the DirecTV and XM-Radio satellites in orbit. This would happen in 2029. The alarmed scientists at NASA even calculated the date: April 13. And, yes, the doomsday near-miss would come on a Friday.

But there was a chance that the asteroid would come so close that Earth's gravity could alter its orbit, nudging it just along a slightly different trajectory, one that would put it on a course to slam into Earth on another go around the sun seven years later. There was an upside, though. This would not be as bad as the asteroid that had hit Earth millions of years ago, wiping out the dinosaurs and 75 percent of all living species at the time. But it would hit with the force of a bomb, somewhere in the Pacific Ocean. Given the size of the rock, and its speed, the astronomers calculated that it would plunge 3 miles down, creating a massive crater and sending a tsunami wave 50 feet high cascading toward California.

Then, fifty seconds after tsunami number one, the water would collapse on itself, filling in the hole the asteroid created on impact. This would touch off a second tsunami. As killer-asteroid-induced tsunamis go, the first wouldn't be so bad, penetrating the shore by only a quarter-mile or so. But it would rip all those fancy coastal houses from their foundation, tear away all those nice restaurants with decks used for sipping cocktails at sunset, and suck them out to sea, where they'd be pulverized in the churn of very angry waters. And then, when the second tsunami hit, it would come crashing back, this time armed with tons of serrated flotsam that would act like sandpaper, wiping away almost everything in its path.

Astronomers named the asteroid Apophis, the Greek name for an Egyptian sun god, a serpent known as the "Lord of Chaos," who symbolized death and darkness.

Thankfully, after studying it for years, the astronomers were able to get a more accurate read of the arc of Apophis's orbit and determined, to their great relief, that though it would still streak closely by in 2029, it would not hit Earth seven years later. So, no worries for now.

Even if it remained an unlikely event, NASA still deemed it important to monitor the cosmos for danger. There's even a specific staff assigned to the task, called the Planetary Defense Coordination Office. It sounds like something out of *Dr. Strangelove*, but it finds and catalogs about 1,500 new near-Earth objects a year, each of which could cause extensive damage if they were to hit Earth.

In the history of the galaxy, the human race has been around for a tiny fraction of time, a mere blink. Life and the rare gift of consciousness did not come with a guarantee that it would continue forever. Asteroids are nature's way of saying, "How's that space program going?" as astronomers like to say.

Musk began thinking seriously about that question, and the probability of an "eventual extinction event," as he called it. The solution: Find another planet to live on. Make humans a multiplanet species, and create a backup hard drive for the human race there, just in case Earth crashes like a faulty computer. The atmosphere on Venus is too acidic. Mercury is too close to the sun. The best bet, he thought, is to colonize Mars.

One night he was driving home from a party on Long Island to New York City with his college friend Adeo Ressi. It was late, and people were asleep in the backseat. But the two friends were deep in an animated discussion.

"We were both interested in space, but we dismissed it as soon as it came up. 'Oh, that's too expensive and complicated.'" Ressi told *Esquire*. "Then two miles would go by. 'Well, how expensive and complicated could it be?' Two more miles. 'It can't be that expensive and complicated.' It kept going on like this, and by the time we made it through the Midtown Tunnel into New York City, we'd basically decided to travel the world to see if something could be done in space."

That night, Musk went back to his hotel, and logged on to NASA's website, looking for the plan to get to Mars.

"Because, of course, there had to be a schedule," he said later. "And I couldn't find it. I thought the problem was me. Because, of course, it must be here somewhere on this website, but just well hidden. And it turned out it wasn't on the website at all. Which was shocking."

It wasn't on the website because there wasn't a schedule.

Although it had achieved enormous success sending robots to the corners of the solar system, NASA's human space program was in a rut. Underfunded, overshadowed by 9/11 and the two wars that followed, space travel had become an afterthought. Since Eugene Cernan became the last man to walk on the moon in 1972, NASA had not sent an astronaut any farther than what's known as low Earth orbit, a few hundred miles up.

Musk, a ravenous reader of science fiction, had expected that by this point in his life there'd be a base on the moon and trips to Mars powered by the robust space program built on the Apollo lunar missions. If in the 1960s, the United States could send a man to the moon in less than a decade, surely there were more great things to come.

He was overcome with what he called a "feeling of dismay."

"I just did not want Apollo to be our high-water mark," he said. "We do not want a future where we tell our children that this was the best we ever did. Growing up, I kept expecting we're going to have a base on the moon, and we're going to have trips to Mars. Instead, we went backwards, and that's a great tragedy."

The more he studied the state of the human space program, the more dismayed he grew. The International Space Station was a marvel, but the way NASA sent astronauts there was, he thought, seriously flawed. The shuttle was mounted on the side of the rocket, like a baby perched on a mother's back, with no way to abort. As a winged spaceplane, it had to re-enter the atmosphere at just the right trajectory, and "even a momentary variance in that can break the vehicle apart. And then, of course, you've got no escape system, so if anything does go wrong, you're toast."

Then there was the cost. NASA was spending billions a year on a lim-ited program flying a handful of times each year, mostly to the space station, which was only some 240 miles away, the same as an Amtrak commute between Washington, DC, and New York. As astronomer Neil deGrasse Tyson put it, the shuttle program "had gone boldly where we had gone hundreds of times before."

Musk had studied physics and economics, and saw all this as a big problem set, a challenge that could be overcome with creative thinking— and his newfound wealth. What separated the Apollo era from now was a Cold War rivalry. But also it was money—and political will. After Apollo, NASA had been routinely starved of funding. Space just didn't capture the public's attention anymore. Shuttle missions had become routine, boring, noteworthy only when there was a tragedy.

Space was still the exclusive domain of governments, but maybe he could pull off something so audacious it would reignite interest in space, get people's attention, and boost funding for NASA.

Musk planned a P. T. Barnum–like stunt he hoped would grab head-lines and reinvigorate interest in space. He'd go out and buy a rocket and launch a greenhouse full of seeds in a nutrient gel that would hydrate upon landing on the surface of Mars. He would create a life support sys-tem on the barren planet and then beam back images of a leafy green plant rising against the lifeless, red landscape. He dubbed the gambit Mars Oasis, and figured he could accomplish it for $15 million to $30 million.

He gathered some of the country's top aerospace minds for a meeting at the Marriott near the Los Angeles airport. Michael Griffin, who later would become NASA's administrator, was there; so was Rob Manning, of NASA's famed Jet Propulsion Laboratory, who served as the chief engineer of Mars Pathfinder, the 23-pound rover that had landed on Mars in 1997. Michael Lembeck, who worked for several aerospace companies, figured Musk's plan would cost far more than Musk suspected. He scribbled the figure $180 million and passed the note to Manning, who had scribbled down his own number.

The two guesses were within $10 million of each other. There was no way Musk could do this on the cheap. Mars was just too difficult—and far. Lembeck had worked in the space industry for a long time, and was skeptical of the whole gambit, he said, because he had seen "a bunch of bright-eyed folks trying to bring the numbers down to commercial prices" while forgetting the axiom that is drilled into every space engineer: "Space is hard."

He calculated that Musk was "at least $100 million off from what was doable." But Musk "didn't want to hear the no word," Lembeck said.

Despite the bad news, Musk came out of the meeting undeterred, vowing to press on.

But the cheapest rocket he could find in the United States was the Delta II, which had cost about $50 million. So, he went to Russia three times in search of a refurbished intercontinental ballistic missile. But that, too, was pricey, and too risky for his taste. Buying a rocket, it turned out, was not such an easy endeavor.

The more Musk studied, the more he realized that there had been very little advancement in rocket technology in the past forty years. The rockets being flown by the Russians and the United States at the early part of the twenty-first century were very similar to those used during the Apollo era. To a self-made Silicon Valley tech entrepreneur, this was stunning.

"The computer that you could have bought in the early '70s would have filled this room, and had less computing power than your cell phone," he said during a speech at Stanford in 2003. "Just about every sector of technology has improved. Why has this not improved? So, I started looking into that."

He gathered a group of engineers and began to meet on Saturdays to figure out "what would be the best way to approach this problem of not just launch cost but of launch reliability."

Musk had read every book he could find on the subject, as Beal had. And he came away convinced that the best way to acquire a rocket was to build it himself, no matter how many times friends told him he was crazy. He shared the banker's zeal of lowering the cost of space travel, and decided he'd try to upend the government-dominated business model that Lockheed Martin, Boeing, Northrop Grumman, and others had been feasting on for years.

On March 14, 2002, Musk incorporated Space Exploration Technologies. Many of his close friends felt the need to stage an intervention. Even Cantrell, one of his earliest advisers, bailed. As impressed as he was with Musk, "I honestly didn't see this whole thing succeeding," he said.

Musk was trying to triumph where even whole countries struggled. At the dawn of the Space Age, from 1957 through 1966, the United States attempted to launch 424 rockets into orbit. Of those, 343 made it successfully—meaning there was a failure rate of nearly 20 percent. During that time, the average number of annual failures—often rockets exploding into angry, red-hot fireballs—was eight. After 1966, there were between one and three annual failures on average, and then one or less a year after 2000. In other words, it took the government's space program nearly five decades to approach anything even close to reliability.

But it was still prone to catastrophic disaster. In early 2003, the Space Shuttle Columbia disintegrated as it reentered Earth's atmosphere, killing all seven astronauts.

A billionaire with no experience in space could not start up a rocket company and a manned space program.

Just ask Andy Beal.

KNOWN AS SPACEX, Musk's new company started out in an old El Segundo factory at 1310 East Grand Avenue, not far from the Los Angeles airport. Musk had sketched out the design of his first creation, a workhorse of a rocket, with a single engine that was purposefully nothing fancy. If others thought of their rockets as racecars, he was happy to compare his to a Honda—utilitarian, reliable, and cheap.

"I would bet you 1,000-to-one that if you bought a Honda Civic that the sucker will not break down in the first year of operation," he told *Fast Company* magazine. "You can have a cheap car that's reliable, and the same applies to rockets." For about $6 million it would be able to launch over 1,000 pounds of payloads, such as satellites, to low Earth orbit, far less than what competitors charged.

It wasn't long before the company's first Falcon 1 rocket was assembled—"Falcon" an homage to the Millennium Falcon from *Star Wars*, "1" denoting the number of first-stage engines it had. But even though Musk had built a rocket in just over a year, he couldn't get anyone at NASA to pay attention.

Washington snubbed Musk just as it had Beal. The establishment—the large contractors, members of Congress, even many in NASA—saw him as just another multimillionaire with a toy space company. A dilettante who couldn't possibly succeed. Few took Musk seriously.

"At the beginning, we had to beg NASA to even pay attention," recalled Lawrence Williams, SpaceX's vice president of strategic relations at the time.

By the end of 2003, Musk decided that if NASA wouldn't come to him, he would go to it. The Federal Aviation Administration (FAA) was preparing to celebrate the hundredth anniversary of the Wright Brothers' first powered flight with a party at the National Air and Space Museum, and Musk decided he'd show up—and bring his new rocket.

For the event, SpaceX loaded the seven-story rocket onto the back of a custom trailer and hauled it cross-country to Washington, DC. With a police escort, it paraded down Independence Avenue, along the National Mall, hallowed ground that had been witness to myriad spectacles, marches, protests. But it had never seen anything quite like this.

As Musk, then thirty-two years old, parked his rocket outside the headquarters of the FAA, tourists who were headed to the National Air and Space Museum stopped to gawk at the streetside exhibit, even in the freezing temperatures. A shiny, white missile that stretched seven stories long, squatting in the real estate usually reserved for hot dog vendors. A cabbie stopped, agog, as the trailer took up an entire lane of traffic—at rush hour. The spectacle was pure Silicon Valley swagger, like an Apple product unveiling, but before Steve Jobs had perfected the art of hyping a new gadget to the masses.

This was Musk's opportunity to show off what his little startup had accomplished—to NASA; to the congressional staffers clamoring for free drinks; to the press, eager for a glimpse—even if it had yet to fly.

But it could fly. It would fly. And its presence on the curb created a stark juxtaposition that was clear and calculated. Inside the museum was NASA's grand past—the lunar lander, the Mercury capsule, the echoes of Apollo enshrined alongside the orphaned dreams it had spawned. Outside was the man who would create a new future—cheap, reliable spaceflight, all with the goal of one day colonizing Mars—a promise as improbable as the young eccentric making it.

He wasn't just selling his rocket, but what it represented—the crazy idea that a small startup could succeed in space. Beal had gotten further than many had thought, and he'd put a nice dent in the wall that kept untraditional players out of the space business. But if Musk was going to avoid Beal's fate, he didn't just have to build reliable rockets—he had to upend the industry's entrenched hierarchy. That would take more than just sound engineering. It would take bravado and guts—a delusion fueled by ego, luck, and an appetite to fight the establishment relentlessly.

The press release announcing the Independence Avenue stunt not only hyped the new rocket as "a major breakthrough in the cost of access to space." It derided the competition as being four times more expensive and far less reliable. SpaceX also exploited the fact that NASA was still grounded ten months after the Space Shuttle Columbia blew up, killing all seven astronauts on board.

"With the grounding of the Space Shuttle creating a backlog in hitchhiker satellite deliveries, there is a great need for new means of access to space," it read, touting the Falcon 1's ability to eventually be reusable.

At the eight p.m. reception, as the NASA officials, congressional staffers, and FAA officials milled about, Musk made his case in a short speech that SpaceX was the answer to a stagnant space industry.

"The history of launch vehicle development has not been very successful; there really hasn't been a success, if you define success as making a significant difference in cost or reliability," he said. "We have a shot with SpaceX, I think, for the first time in a long time."

He invited a small gaggle of press outside, where spotlights highlighted the rocket and a podium had been set up. "We're very proud to debut this vehicle, and to do so here in D.C."

The self-appointed master of ceremonies had even more news to share. This Falcon 1 was just the beginning, he said. The company was already working on a Falcon 5, a far more powerful rocket that would have five first-stage engines instead of just the one. It, too, would disrupt the competition by being far less expensive, he vowed: "It's going to set a new world record for the cost per pound for access to space. That's a huge improvement over anything else."

The Falcon 5 would be big enough to allow SpaceX into the more lucrative market for larger satellites, one dominated by big government contractors. And so this spectacle on the National Mall was more than just a debut for his new rocket. It was a warning shot to such companies as Lockheed Martin and Boeing. Beal had not been able to break their vise grip on the industry. But Musk was armed with a new rocket, and a newly minted fortune that he was ready to burn.

He was coming for them.

3

"Ankle Biter"

A MONTH AFTER MUSK paraded his rocket down Independence Avenue, Sean O'Keefe, then the NASA administrator, had a twenty-one-page report on his desk, detailing SpaceX's capabilities and prospects. "Contains company private data," the title page of the January 29, 2004, document began. "Eyes only. Do not distribute."

Of those in NASA who had heard about SpaceX, few took it seriously. But O'Keefe had grown curious about Musk and his merry band of rocketeers, and wanted to keep an open mind. So, he dispatched one of his lieutenants, Liam Sarsfield, then a high-ranking NASA official in the office of the chief engineer, to California to see whether the company was for real or just another failure in waiting.

Sarsfield was a huge proponent of the commercialization of space and had written a report calling for the agency to rely more heavily on the private sector. Although he'd been looking for a company just like SpaceX to come along, he vowed to give the firm an unbiased assessment and brought a few seasoned colleagues with him. When they walked through the doors of Musk's shop, the quartet became the first NASA officials ever to visit SpaceX's El Segundo headquarters.

It was unlike any rocket company Sarsfield had ever seen. Employees had set up Ping-Pong and air hockey tables and rode around on a Segway. Musk would drive his $1 million McLaren F1 sports car through the hangar door close to his cubicle right onto the factory floor. But the employees were doing real work, building engines and hardware. And as

Sarsfield looked around at the small team that was then just forty-two employees—most engineers and technicians—he saw some faces he recognized from some of the top aerospace companies in the world.

"Aggressive hiring," he noted in his report. "Highly talented, hand-picked team."

Most of all, he was impressed with Musk, who was surprisingly fluent in rocket engineering and understood the science of propulsion and engine design. Musk was intense, preternaturally focused, and extremely determined.

"This was not the kind of guy who was going to accept failure," Sarsfield remembered thinking.

Throughout the day, as Musk showed off mockups of the Falcon 1 and Falcon 5, the engine designs, and plans to build a spacecraft capable of flying humans, Musk peppered Sarsfield with questions. He wanted to know what was going on within NASA. And how a company like his would be perceived. He asked tons of highly technical questions, including a detailed discussion about "base heating," the heat radiating out from the exhaust going back up into the rocket's engine compartment—a particular problem with rockets that have clusters of engines next to one another, as Musk was planning to build.

Now that he had a friend inside of NASA, Musk kept up with the questions in the weeks after Sarsfield's visit, firing off "a nonstop torrent of e-mails" and texts, Sarsfield said. Musk jokingly warned that texting was a "core competency."

"He sends texts in a constant flow," Sarsfield recalled. "I found him to be consumed by whatever was in front of him and anxious to solve problems. This, combined with a tendency to work eighteen hours a day, is a sign of someone driven to succeed."

Musk was particularly interested in the docking adapter of the International Space Station, the port where the spacecraft his team was designing would dock. He wanted to know the dimensions, the locking pin design, even the bolt pattern of the hatch. The more documents Sarsfield sent, the more questions Musk had.

"Most of us struggle with fear," Sarsfield said. "We fret about this and that and generally dread looking dumb. I found Elon fearless in this regard. He's not afraid to ask a question that proves he doesn't understand something. . . .

"I really enjoyed the way he would pore over problems anxious to absorb every detail. To my mind, someone that clearly committed deserves all the support and help you can give him."

Sarsfield told his bosses at NASA that Musk could very well make it. In his report to O'Keefe, he predicted that "Falcon 1 will be successful with initial launch series"—though not on its first launch. And he concluded that "SpaceX presents good products and solid potential—NASA investment in this venture is well warranted."

But inside the agency headquarters, there were more skeptics than believers.

"I would say ninety-five percent of the people at NASA thought he would most certainly fail," Sarsfield said. "I'd say to them, 'I hear you, and Elon will have bumps along the road. But I guarantee you this guy is not going to fail.'"

A MONTH AFTER Sarsfield issued his report, Musk e-mailed him again. But this time the tone was different. NASA had just awarded a $227 million sole-source contract to another commercial space company, Kistler Aerospace. Musk wanted to know why SpaceX—or any other company—hadn't been allowed to bid on it.

Musk may have had his newly minted millions, but Kistler had considerable pull in NASA and Washington, DC. The company was led by George Mueller, an aerospace legend who had headed the Office of Manned Spaceflight during the Apollo era. He was considered something of a hero at NASA, who along with Wernher von Braun had helped the country fulfill John F. Kennedy's goal of sending a man to the moon by the end of the 1960s.

Mueller later helped design the first space station and was considered the "father of the space shuttle." In 1971, the year Musk was born, Richard Nixon had awarded Mueller the National Medal of Science in a ceremony in the East Room, "for his many individual contributions to the design of the Apollo system."

After his government career, Mueller had turned to the private sector, serving as a senior vice president at General Dynamics before taking over as chief executive at Kistler, a small startup. The young company was in trouble, and had filed for bankruptcy in 2003, owing $600 million to creditors. The NASA contract would help it stay afloat.

Musk was incensed, and felt that the contract was unfair, if not illegal. Kistler was hurting, Sarsfield wrote to him, noting that its executive had long ties to NASA and that "I worry that Kistler's financial arrangements are shaky (a conservative word), but the money is pocket change when you look at how much we blow through per annum."

SpaceX shouldn't worry, Sarsfield wrote; there would be other contracts coming. But that only made Musk angrier, and more determined. Like Andy Beal, he felt that NASA's role wasn't to prop up chosen companies. Competition would promote better and safer technologies, at lower costs. This was an old-boys network, and he wanted in—or to smash it.

Musk took his complaint to top NASA officials, and in a meeting at NASA headquarters in Washington, threatened to file a legal challenge over the no-bid contract with the Government Accountability Office (GAO). His colleagues warned him that it was not a smart business decision to threaten an agency that could make or break SpaceX. At the meeting, NASA officials intimated that a lawsuit would not be in SpaceX's best interests. If Musk sued, they might never work with him.

"I was told by everyone that you do not sue NASA," Musk recalled. "I was told the odds of winning a protest were less than ten percent, and you don't sue your potential future customer. I was like, look, 'This is messed up. This should have been a competed contract, and it wasn't.'"

It was a simple matter of right and wrong, though that logic didn't always appease the executives who'd have to be the ones to work with NASA.

"Being the customer relationship person, I was always very worried about that," said Gwynne Shotwell, who would become SpaceX's president and chief operating officer. "But Elon fights for the right thing. And he says if people are going to get offended by you fighting for the right thing, then they are going to get offended."

From the beginning, SpaceX's mantra was to "set audacious, nearly impossible goals and don't get dissuaded. Head down. Plow through the line. That's very SpaceX," she said. "That's kind of the deal."

Musk exuded swagger and confidence, and it spread to his employees. "SpaceX is a place where you get to be mouthy," Shotwell said. "You get to express your opinion. You get to push really hard."

Still, Lawrence Williams, one of the few people SpaceX had in Washington to work government relations, got the message and emerged shaken from the meeting at NASA. He had spent most of his career in Washington, and had worked on the Hill as an aide on the House Science, Space, and Technology Committee. The message from NASA was clear, he said: "Elon, if you pursue this, you will lose and likely never do business with NASA."

But Musk was unfazed. "He didn't even blink," Williams said. "Despite everyone's stern warnings, Elon didn't hesitate to sue the entity he wanted as our customer more than any. In my twenty-plus years in Washington, I never witnessed anyone with more conviction and confidence, who never hesitated to risk it all for something he believed."

Head down. Plow through the line.

In its suit, SpaceX even included Sarsfield's e-mail as evidence that the contract was to help save Kistler. "This goes to show you the way Elon plays ball," Williams said. "He files as a part of the government contract protest the e-mail from Liam Sarsfield, who was then probably our only friend at NASA, saying that 'this was a life preserver to Kistler—and don't worry; we'll try to do something to help you out down the road.'"

SpaceX got support from Citizens Against Government Waste, a good government nonprofit whose president, Tom Schatz, said Musk caught NASA trying "to pull a fast one, bypassing full and open competition requirements by doing a sloppy job of assessing the qualifications of other applicants." He said that an "unwarranted sole-source contract that stinks of a kickback to former employees is a bad omen for NASA's privatization efforts."

Musk even brought his fight to Capitol Hill. He'd been invited to testify before a Senate committee in May 2004 about the future of space launch vehicles, and the role private industry might play. But, blunt as always, he planned to use the audience to his advantage. Musk's prepared testimony started out going for the jugular, reminding Congress of its long track record of funding flops.

"The past few decades have been a dark age for development of a new human space transportation system," he said. "One multibillion-dollar government program after another has failed. In fact, they have failed even to reach the launchpad, let alone get to space. . . .

"The reaction of the public has been to care less and less about space, an apathy not intrinsic to a nation of explorers, but born of poor progress, of being disappointed time and again. When America landed on the moon, I believe we made a promise and gave people a dream. It seemed then that, given the normal course of technological evolution, someone who was not a billionaire, not an astronaut made of the 'Right Stuff,' but just a normal person, might one day see Earth from space."

He went on to propose three ways Congress might help that happen: create more prizes that industry might compete for; focus on vehicles that would lower the cost to space; and ensure fairness in government contracting.

This was where he wanted to shine the spotlight on his fight over the Kistler contract and bring it to the attention of Congress. He complained that SpaceX and others "were denied the opportunity to compete on a level playing field to best serve the American taxpayer." And he took a dig at Kistler, writing that its award "is mystifying given that the company has been bankrupt since July of last year, demonstrating less than stellar business execution (if a pun is permitted)."

But before he could read his statement to the committee, Sen. John Breaux, a Democrat from Louisiana, raised an objection. He did not want Musk litigating his bid protest at a Senate hearing, saying it was "patently unfair to not have the other side present it at the same forum, dealing with something that's under a contract dispute."

It didn't matter. Blunt as always, Musk had made his point. And his lawyers had laid out a convincing case that the contract should have never been awarded without competition. The GAO, which oversaw the protest, forced NASA to withdraw the contract. SpaceX had won. NASA would later open up another contract, and this time SpaceX could compete.

"That was a huge upset—literally imagine, like, a ten-to-one-odds underdog winning," Musk said years later. "People did not expect this. It blew everyone's mind that the GAO sided with SpaceX. There are some brave and honest and true people at the GAO. They're great because they were under such pressure to rule against us. Massive pressure. But that victory with the GAO was important for the future of SpaceX."

MUSK MAY HAVE scored a momentous legal victory over the Kistler contract, but it didn't win him any friends. If anything, the Washington establishment he was trying to woo at the Air and Space Museum event turned even colder.

It didn't help that SpaceX got into a fight in early 2004 with Northrop Grumman. Northrop had been chosen by the Pentagon to oversee SpaceX's rocket development work. The Pentagon had grown interested in this spunky little company and was keenly interested in harnessing new technology that could help it launch satellites quickly.

One top air force official called Musk a "pathfinder" and said, "We need him to be successful."

But if the Department of Defense was going to trust this new company and its unconventional way of doing business, it wanted deep insight into its manufacturing processes, its workforce and engine designs. Given budget constraints, the Pentagon didn't have the manpower, however, so it outsourced the oversight to some of its most trusted contractors.

That's how a team of Northrop Grumman engineers found themselves stationed at SpaceX's facility in El Segundo. The only problem was, Northrop was a competitor that also built rocket parts for the Pentagon. There was an inherent conflict of interest. The Pentagon claimed it was doing its best to mitigate the problem—companies like Northrop were supposed to sequester the employees who were embedded with rivals to make sure they didn't interact with their colleagues working on similar projects.

"We do everything under the sun" to make sure supervising companies don't walk away with the secrets of the companies they are supervising, a Pentagon official told the *Wall Street Journal*.

The arrangement was tenuous at best. In January 2004, Gwynne Shotwell, then SpaceX's head of business development, was concerned that the Northrop team was using its access to benefit Northrop. She burst into a meeting and demanded to know whether any of the Northrop team was also involved in developing Northrop's engines. Five of the eight Northrop teammates raised their hand, the *Journal* reported.

The air force replaced the Northrop team with a group of engineers from another company, the Aerospace Corporation. But the damage was done. Northrop fired the first salvo, filing a lawsuit in May, accusing SpaceX of using its rocket-engine designs. Musk had hired Thomas Mueller, SpaceX's chief propulsion engineer, away from a Northrop subsidiary. Northrop alleged he had brought over inside knowledge about its programs. Northrop also alleged that its rival had many of its internal documents, stamped "proprietary."

SpaceX denied the charge, and fired back a month later with its own lawsuit. It accused Northrop of abusing its supervisory position as part of a "surreptitious" attempt to engage in corporate espionage and alleged that Northrop had failed to "preserve, to protect and to not misuse proprietary information."

Ultimately, the companies settled and dropped their lawsuits. The legal fight was costly and a distraction, Musk said. But it was important for his little startup to stand up to a company he felt was bullying it. "Northrop wasn't expecting us to fight as hard as we did," he said.

As a kid in South Africa, Musk had been bullied relentlessly. Once he'd even been thrown down a flight of stairs and beaten so badly he'd ended up in the hospital. At the time, he'd retreated into books and computers, reading and playing video games for hours at a time.

If SpaceX was going to be successful, showmanship on the National Mall was not going to be enough. He was going to have to battle his way in, facing strength with strength. This time he would punch back.

IN THE YEARS after the September 11 attacks, the Pentagon and intelligence agencies were relying on space more than ever before. The satellites in orbit were playing an increasingly important role, providing secure communications to troops on the ground, often in remote locations. They also beamed GPS guidance to weapons, precision guided munitions, and the drones that were beginning to swarm the battlefields in Iraq and Afghanistan.

Getting those satellites into space reliably was important—and big business. While NASA may have had the prestige and pedigree, the real money in space at the time was to be made from the Pentagon. For years, the national security space market had been dominated by Boeing and Lockheed Martin. By law, the Pentagon was required to have "assured access to space," meaning it needed to have at least two rockets certified to launch military and intelligence satellites. If one suffered a failure, there would be a backup. In theory, the companies would also compete against each other—keeping the prices down.

In 1998, the Pentagon held a competition for launches worth hundreds of millions of dollars. Boeing came out ahead, winning nineteen launches, while Lockheed won just nine. The decision was a stunning blow for Lockheed, which had been the Pentagon's preeminent supplier. But federal investigators would later determine that Boeing had illegally acquired thousands of pages of Lockheed's internal, proprietary data that gave Boeing an enormous advantage in the competition.

The scandal rocked Washington, and the air force suspended the aerospace giant, hitting it with what the *Wall Street Journal* called the "stiffest punishment imposed on any major Pentagon contractor in decades." It took away $1 billion worth of business, transferred seven launches to Lockheed, and also awarded it three more without competition.

"I've never heard of a case of this scale," air force secretary Peter Teets said at the time.

But by 2005, the companies made up—or were forced to. There simply wasn't enough business to sustain them both, they said, and so they announced plans to consolidate their space launch businesses into a single company. The merger would create a behemoth unlike anything the Pentagon had ever seen. Called the United Launch Alliance, it paired the Pentagon's top two contractors, giving them a monopoly on billions of dollars of its business.

By banding together, the companies had extraordinary leverage over the Pentagon, which didn't have anyone else to turn to for satellite launches. Defense officials not only approved the merger, but agreed to also provide the combined company with an additional contract, this one to cover any of the company's overhead costs, worth hundreds of millions of dollars.

By now, Musk had sued over Kistler and had battled Northrop Grumman, and he was not going to let the formation of a monopoly go without a fight. The consequences of the merger for SpaceX could be dire: the Alliance was on the verge of locking up a huge contract that would have given it a monopoly through at least 2011. So, SpaceX filed suit in October 2005, alleging that the companies used "strong arm tactics" to force the Pentagon to approve the merger and then hand it exclusive contracts that have "destroyed any pretense of competition in the sale of [rocket launches] to the government."

"SpaceX poses a significant threat to Boeing and Lockheed Martin's dominant position," the suit stated. "It has developed new technologies and a new business model that will allow it to reduce dramatically the cost of access to space and increase the reliability of launch vehicles. The rockets being developed by SpaceX will perform better, and will be much less expensive, than those offered by Boeing or Lockheed Martin."

It accused the firms of staging "a boycott" to force the Pentagon's hand, which would then "exclude all other competitors, including SpaceX."

For a company that had yet to launch a rocket, it was quite a lawsuit. Lockheed and Boeing denied the claims, and SpaceX's suit was dismissed. The Alliance was born, and would for a decade hold a monopoly on national security launches, winning billions of dollars from the Pentagon.

Musk vowed that he would continue to fight. But Boeing and Lockheed didn't seem particularly worried about Musk or his hard-charging upstart.

Williams, one of SpaceX's top Washington operatives, recalled that Lockheed's head lobbyist "used to refer to us on the Hill as an 'ankle biter.'" Another company derided SpaceX's rockets as being made "out of bicycle parts."

Backed by Musk's vision and millions, SpaceX was certainly brash—but it was unproven. In the space business, talk was cheap. And the companies didn't limit their derision to private Capitol Hill meetings.

"SpaceX needs to prove themselves, and thus far they have been unable to demonstrate that they are a competitor," a Lockheed spokesman told the *New York Times*.

Boeing was just as dismissive: "Launching into space is an extremely challenging and complex business. For SpaceX to be considered a potential competitor they need to have a launch."

DOWN IN MCGREGOR, they were working on it. Musk had assembled a small crew to work on the engine test firings on the Texas plain. It was unlike anything he had ever been involved with. Failed software led to 404 error messages, or a crashed hard drive. These miscalculations set off deafening, window-rattling explosions that sent the cows on nearby pastures running. The failures were so frequent, and loud, that the SpaceX employees, stuck in McGregor with little to do for entertainment, set up a "cow cam" to record their scattering like birds.

SpaceX soon outgrew the modest footprint it had rented from the city, and then rented more. Soon it grew from 197 acres to 256. Then 631, then to more than 1,000 acres. Finally, SpaceX took over virtually the entire facility, some 4,000 acres, an amphitheater for the growing cacophony Musk was producing 4 miles west of the high school. More room for bigger engines. More smoke and fire. More noise.

While the company's headquarters were out in California, "I tell folks if they really want to see the exciting stuff, they should come to Texas," Musk said. "That's where we light the fires and where most of our advanced engines are."

At a nearby state park, rangers put up signs warning visitors that the roar they would occasionally hear was not, in fact, the end of the world, just SpaceX firing off another engine. "While at Mother Neff State Park you may hear a 'thunderous' sound throughout the day or night. If the skies are clear, then no need for concern. The continual thunderous sound is coming from the SpaceX rocket research and development center 6 miles north of the park."

Musk's engineers in Central Texas were making as much of a ruckus as the lawyers he had hired to take on the establishment in Washington. Musk was putting on quite a show—the high-profile lawsuits, the police-escorted parade down Independence Avenue, the congressional hearing, and the crescendo of the fiery rocket tests—hosted by Silicon Valley's newest boy wonder.

The company had yet to fly, and it wasn't clear whether the company would ever amount to anything. Early on, Musk pegged his chance of success at 10 percent. But his full-throttle performance was at least beginning to achieve one of his early goals—to reinvigorate interest in space.

If nothing else, he was getting people's attention.

NEARLY 500 MILES to the west, on the land he had purchased clandestinely, Bezos was quietly building his own rocket company. Obsessed with secrecy, he was as quiet and slow as Musk was loud and fast. While Musk thrust his rocket under the spotlight on Independence Avenue, Bezos kept his company's work hidden.

Musk had heard that Bezos had started a rocket company as well, and was curious to learn more. "I think he was concerned that Amazon investors would think he sort of has this weird distraction," Musk later recalled. The pair had dinner in about 2004, Musk said.

"We talked about rocket architectures," Musk later recalled. "It was very clear technically he was barking up the wrong tree, and I tried to give the best advice I could. . . . Some of the engine architectures they were pursuing were the wrong evolutionary path."

Some of the ideas that Bezos proposed, SpaceX had already been tested, Musk said. "Dude, we tried that and that turned out to be really dumb, so I'm telling you don't do the dumb thing we did," he recalled saying. "I actually did my best to give good advice, which he largely ignored."

Unlike Musk, Bezos wasn't in a rush. He was happy to experiment and fail, to try new ideas, even if they had been tried before and went nowhere. Bezos had vast amounts of patience. This was a man who was, after all, building a ten-thousand-year clock inside of a mountain on his West Texas property that would be "a symbol, an icon for long-term thinking," he had written. It would have "a century hand [that] advances once every 100 years, and the cuckoo comes out in the millennium" as it kept time for ten thousand years.

In addition to its logo, a feather, Blue Origin had a coat of arms that it would eventually showcase on the wall of its headquarters in Kent, Washington, outside Seattle. It was an involved piece of art, loaded with symbolism, from Earth to the stars to the velocities needed to reach various altitudes in space. There was also a winged hourglass, representing human mortality.

"Time is fleeting," Bezos said once on a tour of the facility. Despite the seemingly plodding steps he would take, he had a sense of urgency and a direction. Only, as he said, "you get there faster when you take one step at a time."

The company's motto was "Gradatim Ferociter" (step by step, ferociously). The phrase appeared across the bottom of the coat of arms. But perhaps none of its symbols was more important than a pair of turtles reaching up toward the stars—an homage to the winner of the race between the tortoise and the hare.

The turtle was Blue Origin's mascot, the embodiment of another of Bezos's favorite sayings, one derived from US Navy SEAL training: "Slow is smooth and smooth is fast." It was the opposite of SpaceX's "Head down. Plow through the line."

Now Musk and Bezos were playing the parts in a modern version of Aesop's fable. The hare had burst forward, dashing ahead, kicking up a tumultuous plume of dust, while the tortoise creaked slowly, uttering softly in an I-think-I-can cadence:

Slow is smooth and smooth is fast. Slow is smooth and smooth is fast. Slow is smooth and smooth is fast.

4

"Somewhere Else Entirely"

O N OCTOBER 9, 1957, five days after the Soviet Union launched Sputnik, the first ever satellite into orbit, President Dwight D. Eisenhower stood before an unusually hostile press corps in the Old Executive Office Building. For days, his administration had downplayed the significance of the Soviets' feat. But by now the country had grown increasingly alarmed, and he needed to respond.

Eisenhower entered the room at 10:31 a.m., and decided to get right to it, asking, "Do you have any questions for me?"

The very first question he faced, from United Press International, was blunter than he was used to: "Mr. President, Russia has launched an Earth satellite. They also claim to have had a successful firing of an intercontinental ballistic missile, none of which this country has done. I ask you, sir, what are we going to do about it?"

In the midst of the Cold War, the Soviets' launches were seen as acts of aggression, expressions of military superiority. In a memo to the White House, C. D. Jackson, a former special assistant to the president who had served in the Office of Strategic Services, wrote that it was "an overwhelming important event—against our side. . . . This will be the first time they have achieved a big scientific jump on us, ostensibly for peaceful scientific purposes, yet with tremendous military overtones. Up to now, it has generally been the other way around."

If the Soviet Union could put a satellite into orbit, it's hold the ultimate high ground and could, many feared, rain down missiles on American cities from space. *Life* magazine compared Sputnik to the shots fired at Lexington and Concord and urged the country to "respond as the Minutemen had done then." Then Texas senator Lyndon Johnson fretted that "soon they will be dropping bombs on us from space like kids dropping rocks onto cars from freeway overpasses."

Eisenhower's answer to the reporter's pointed question was, in essence, that the country was working on it. The real response to the Soviets would come a few months later, when during his 1958 State of the Union address, he talked about the creation of a new agency within the Defense Department that would have "single control in some of our most advanced development projects." This agency would be in charge of "antimissile and satellite technology" at a time when "some of the important new weapons which technology has produced do not fit into any existing service pattern."

The Soviets' launch of Sputnik opened a new frontier—space—one that "creates new difficulties, reminiscent of those attending the advent of the airplane a half century ago," he said.

The new organization would be called the Advanced Research Projects Agency (ARPA). Born from what the secretive agency now calls the "traumatic experience of technological surprise," ARPA would be a sort of elite special force within the Pentagon made of its best and brightest scientists and engineers. But because it would transcend the traditional services—the army, navy, air force—many in the defense establishment looked askance at it.

Eisenhower didn't care. To keep up with the Soviets, the nation needed to move past "harmful service rivalries," he said.

Some of the top brass in the Pentagon were charged with single-handedly picking top talent for ARPA, renamed DARPA in 1972—the "D" for "Defense." Successful candidates would have to not only be smart and efficient, but they'd also have to be morally strong and confident, able to stand up to generals and admirals that might resent their very presence and consider them outsiders.

They were encouraged to push boundaries, and create new, futuristic technologies that aimed at keeping the nation several steps ahead.

"In the 1960s you could do really any damn thing you wanted, as long as it wasn't against the law or immoral," Charles Herzfeld, who directed ARPA from 1965 to 1967, told the *Los Angeles Times*.

WILFRED MCNEIL, THE Pentagon's comptroller, helped recruit top talent to help run the agency. One of his top choices was Lawrence Preston Gise, a stolid and principled former navy lieutenant commander. Born in Texas, Gise had served during World War II, and service records show he was assigned to the USS *Neunzer*, a destroyer, and then to various administrative jobs. He also served as an assistant director at the Atomic Energy Commission, starting in 1949, and was promoted to assistant director in 1955.

By the height of the Cold War, Gise found himself in the middle of an agency that was developing the hydrogen bomb. As a young employee, he had participated in a secret meeting in 1950 to discuss the development of the bomb with some of the agency's top officials, including its then-chairman Gordon Dean.

Gise was intrigued by the possibilities of ARPA, and what it represented at the dawn of the Space Age. But he was also aware that political pressure was mounting against its formation. With a family to support, he hedged his bets, making sure he would have a landing spot, just in case this experimental agency didn't work out.

"So the agency was controversial even before it was formed," Gise said in a 1975 history of ARPA. "My deal with McNeil was I would come over and handle the administrative side of the business with the assurance that if the agency went up in blue smoke that he would absorb me in his immediate office, and he had a job set up for that purpose. But it was that tenuous back in those days."

Gise was well respected by the agency's director, Roy Johnson, who had left a high-paying job as an executive at General Electric for the post at ARPA. His goal was to ensure the country caught up and passed the Soviets, focusing much of his energy on space.

"Johnson believed that he had personally been given unlimited authority by the Secretary to produce results," according to the ARPA history. "He really thought that he was supposed to be the czar of the space program. . . . Johnson perceived that ARPA's job was to put up satellites. The space program became his principal interest."

After three years at ARPA, Gise was lured back to the Atomic Energy Commission, which offered him a job in top management. But he continued to work alongside the agency, collaborating on an endeavor known as the Vela Project, which was designed to detect nuclear explosions from

space through a high-altitude satellite system. In a message to his colleagues, Gise reported that "ARPA is implementing on a very urgent basis
a program to establish its capability for detection of Argus effects"—an
apparent reference to Operation Argus, three high-altitude nuclear test
explosions over the South Atlantic Ocean in 1958.

Gise would continue to serve at the Atomic Energy Commission until
1968, when he wanted to close a factory that politicians wanted to keep
open. The politicians prevailed, and Gise retired to his ranch in South
Texas.

He was young, just fifty-three years old. But he was looking forward
to life on the ranch. Plus, he had a young grandson to tend to, a remarkable little boy with big ears and a wide smile, who shared his middle
name:

Jeffrey Preston Bezos.

IN ADDITION TO being a top defense official, Gise was a dedicated family
man, who looked after Bezos's mother, Jackie, after she got pregnant with
Jeff. She was just seventeen when Jeff was born and Jackie married his father, Ted Jorgensen. Gise supported them, flying the couple to Mexico to
get married, then hosting another ceremony at their house.

He paid his son-in-law's tuition at the University of New Mexico, but
Jorgensen dropped out. Then, Gise tried to land him a police department
job, but that didn't work out, either.

Neither did the marriage. The young parents soon got divorced, and
Jackie took her son and moved back in with her parents in Albuquerque.

Jackie got a job at the bank of New Mexico and met a hardworking
man there named Miguel Bezos, known as Mike, who had fled Cuba
shortly before the Cuban missile crisis. They fell in love and got married
when Jeff was four. Mike Bezos adopted him and raised him as his own.
Gise made Jorgensen promise to stay away.

"I've never been curious about him," Bezos told *Time* about his biological father. "My real father is the guy who raised me."

Bezos's passion for space started when he was five years old on July 20,
1969, when Neil Armstrong and Buzz Aldrin landed on the moon. As
young as he was, he could tell he was watching something historic.

"It really was a seminal moment for me," he said. "I remember watching it on our living room TV, and the excitement of my parents and my

grandparents. Little kids can pick up that kind of excitement. They know something extraordinary is happening. That definitely became a passion of mine."

The family lived in New Mexico and Texas and later in Florida. But after school got out for the year, Bezos was shipped off to the ranch, where he spent every summer from the ages of four to sixteen.

Located in Cotulla, a small town about 90 miles south of San Antonio, it was rural and isolated, a place where Bezos learned the value of self-reliance from his grandfather. "Pop," as Bezos called him, was patient and gentle and taught his grandson to live a rancher's life, fixing windmills and laying pipe. On the ranch, Bezos learned to vaccinate and castrate cattle, and brand them with the ranch's Lazy G logo. And when the D-6 Caterpillar bulldozer broke, Pop and his eager grandson built a crane to lift the huge gears out.

It was, Bezos recalled in an interview with the Academy of Achievement, a nonprofit, "an incredible, incredible experience. Ranchers, and anybody I think who works in rural areas, they learn how to be very self-reliant. And whether they're farmers, whatever it is they're doing, they have to rely on themselves for a lot of things."

Bezos spent a lot of time with his grandfather, who he said was "always incredibly respectful of me even when I was a little kid. And would entertain long conversations with me about technology and space and anything I was interested in."

His grandparents were also members of a "Caravan Club," striking out across the United States and Canada, sometimes taking their inquisitive grandson along for the ride.

"We'd hitch up the Airstream trailer to my grandfather's car, and off we'd go, in a line with 300 other Airstream adventurers," Bezos said in 2010 during a graduation speech at Princeton. "I loved and worshipped my grandparents, and I really looked forward to these trips."

He recalled one trip, when he was about ten, "rolling around in the big bench seat in the back of the car." Pop Gise was at the wheel. Bezos's grandmother, Mattie, was beside him, smoking as she always did on these trips, filling the car with a smell that Bezos couldn't stand.

Bezos remembered an antismoking campaign advertisement he'd recently heard about the perils of smoking that said that every puff takes about two minutes off your life. Even at age ten, Bezos loved coming up with math calculations in his head, estimating how far they'd be able to travel on a tank of gas, what they'd spend at the grocery store. And with his grandmother puffing away on the front passenger seat, and an open road with little else to occupy his expansive mind, Bezos decided to do the math.

"I estimated the number of cigarettes per day, estimated the number of puffs per cigarette and so on," he told the Princeton graduates. "When I was satisfied that I'd come up with a reasonable number, I poked my head into the front of the car, tapped my grandmother on the shoulder, and proudly proclaimed, 'At two minutes per puff, you've taken nine years off your life!'"

He expected his grandparents would be awed by his precociousness. "Jeff, you're so smart. You had to have made some tricky estimates, figure out the number of minutes in a year and do some division."

But instead the car was silent, except for the sound of his grandmother's sobs.

"While my grandmother sat crying, my grandfather, who had been driving in silence, pulled over onto the shoulder of the highway," Bezos said. "He got out of the car and came around and opened my door and waited for me to follow.

"Was I in trouble? My grandfather was a highly intelligent, quiet man. He had never said a harsh word to me, and maybe this was to be the first time? Or maybe he would ask that I get back in the car and apologize to my grandmother. I had no experience in this realm with my grandparents and no way to gauge what the consequences might be.

"We stopped beside the trailer. My grandfather looked at me, and after a bit of silence, he gently and calmly said, 'Jeff, one day you'll understand that it's harder to be kind than clever.'"

Bezos spent his summers on the ranch, even though the stifling heat would often drive them to huddle indoors. They'd watch soap operas. *Days of Our Lives* was a favorite. His grandparents encouraged playing board games and reading, and Bezos discovered that the county library, which was not much larger than a one-room schoolhouse, had an extensive science fiction collection that had been donated by a town resident.

The library "had maybe a few hundred science fiction novels. All of the classics," Bezos recalled. "There was a whole shelf of them there, and over several summers I worked my way through that collection."

The visits to the library "started a love affair for me with people like [Robert] Heinlein and [Isaac] Asimov, and all the well-known science fiction authors that persists to this day."

The ranch, where the big sky opened up dark and deep, was an ideal place for a starry-eyed kid who dreamed of one day becoming an astronaut to indulge his science fiction fantasies.

AT HOME, BEZOS spent a lot of time watching *Star Trek*, his favorite. But in the fourth grade, when he was nine years old, he figured out how to play a Star Trek game on a computer at school. It was 1974, before the advent of the personal computer; his elementary school had one mainframe with a teletype connected to an acoustic modem. Not that anyone at the school knew how to use it. "But there was a stack of manuals, and me and a couple of other kids stayed after class and learned how to program this thing," he recalled, eventually figuring out that it had been preprogrammed with the Star Trek game.

"And from that day forward all we did was play Star Trek," he said.

Later, he'd even named his dog Kamala, after the character on *Star Trek*.

By the time he got to high school, Bezos's passion for space merged with his prodigious intellect and curiosity. In high school, he wrote an essay titled "The Effect of Zero Gravity on the Aging Rate of the Common Housefly" that won him a trip to NASA's Marshall Space Flight Center in Huntsville, Alabama.

His idea was to test how the weightless environment in space reduced stress on the body's systems. Bezos thought to start out with a creature with a very short life span—the common housefly—to test whether you could see any biological changes in a short amount of time aboard the space shuttle, compared with a control group of flies kept on the ground.

He was a finalist, not a winner, so NASA never did fly his experiment to space. But he and his physics professor got to spend a couple of days at Marshall. It didn't have the cachet of the Kennedy Space Center in Florida, where the astronauts launched into space, or the Johnson Space Center in Houston, where they trained. Rather, it was where NASA built its rockets, the home to many of its most accomplished engineers and sharpest minds.

When he was "really young I wanted to be an astronaut," he said. "I went through many phases and wanted to do many different things. I wanted to be an archaeologist—and this was pre–Indiana Jones. I didn't have Indy on the mind. Many kids sort of have ideas of what you want to do and so on and the thing that never went away was my fascination with space. And then I realized I didn't want to be an astronaut. I was really more interested in the engineering side of it."

Marshall was the perfect place, then, for someone as eager and inquisitive as Bezos, an inveterate tinkerer who said that "our garage was basically science fair central." His mother joked that she was singlehandedly keeping Radio Shack in business by buying him parts for the projects he was building in the garage.

"Will you please get your parts list straight before we go?" she'd chastise him. "I can't handle more than one trip to Radio Shack per day." He had such concentration that as a toddler in Montessori School, his teachers would have to pick him up—in his chair—to keep him moving from task to task.

Under the leadership of Wernher von Braun, one of the fathers of rocketry, the Marshall Space Flight Center was where NASA built the F-1 rocket engine that powered the Saturn V rocket to the moon. The engines were massive, towering nearly 20 feet high and more than 12 feet wide, and weighed 18,000 pounds. They were monuments to engineering that in a cluster of five burned through liquid oxygen and kerosene propellants at more than 15 metric tons per second. Bezos was in awe of their power, 1.5 million pounds of thrust, and the intricate mechanics that went into the most powerful liquid fuel engine ever to fly.

The trip only strengthened his enthusiasm for space. If he'd fed himself a steady diet of science fiction fantasy in the books he'd devoured as a kid, now he was exposed to actual hardware—the equipment that made space dreams real.

"He raved about it," said Joshua Weinstein, a friend from high school.

At Miami Palmetto High, their class was full of exceptionally talented kids who would go on to great things. "It was a hard class to stand out in, but Jeff did," Weinstein said. Bezos did so academically, graduating first in his class. He enjoyed school, was a voracious reader, and wanted to please his teachers.

"I was very difficult to punish for my parents because they would send me to my room, and I was always happy to go to my room because I would just read," he said. Once he was laughing too loudly, and lost his library privileges, which he said "was really inconvenient for me."

But he also had a cunning, rebellious wit, laying booby traps across the house. "I think I occasionally worried my parents that they were going to open the door one day and have 30 pounds of nails drop on their heads or something," he said.

Weinstein remembered how Bezos had been carrying on—loudly—when a stern teacher named Bill Henderson chastised him.

"Mr. Bezos!" he bellowed.

"It's Jeff to you," Bezos bellowed back. "Only my friends call me Mr. Bezos!"

The class broke out into laughter—and the teacher did, too.

As valedictorian, his speech at graduation was about space. For even a brilliant eighteen-year-old, it was a precocious glimpse into the future. He talked about plans to colonize space, to build habitats like space hotels, and the day when millions of people were living among the stars. Earth had limited resources and so his idea was to get humanity off its surface, into space so as to protect the planet. He concluded by saying, "Space, the final frontier, meet me there!"

"The whole idea is to preserve the Earth," he told the *Miami Herald* at the time, saying it should be designated as a national park.

Space, and humanity's future in it, was something he had been thinking and reading about for some time.

"He said the future of mankind is not on this planet, because we might be struck by something, and we better have a spaceship out there," Rudolph Werner, the father of Bezos's high school girlfriend, told *Wired* magazine.

THE TALK OF space hotels, with amusement parks, yachts, and colonies for 2 to 3 million people in orbit—all a way to help preserve Earth—these were not the elements of a normal high school graduation speech. They were the science fiction–fueled musings of one of "Gerry's kids"—the devotees of Gerard O'Neill, a Princeton physics professor and space visionary, whose book *The High Frontier* became a manifesto for such enthusiasts as Bezos.

Years before Bezos gave his graduation speech, the professor had gained widespread attention for his plans to colonize space. In 1974, the

New York Times had covered a conference O'Neill hosted at Princeton, gathering some of the nation's leading engineers. The article came in the post-Apollo hangover, when NASA's budget had been gutted and interest in space waned. But the piece carried a sensational headline: "Proposal for Human Colonies in Space Is Hailed by Scientists as Feasible Now." It helped put O'Neill, and his ideas, on the map.

"The initial goal would be construction of a small colony of about 2,000 people at a site, along with the orbital path of the moon, known as the L-5 libration point," the article read. It quoted O'Neill as saying that most "dirty" industry could move into space, allowing the preservation of Earth, which, O'Neill said, would become a "worldwide park, a beautiful place to visit for vacation"—words that Bezos would echo in his graduation speech years later.

In 1977, the year *The High Frontier* was published, O'Neill appeared on the *Tonight Show with Johnny Carson* to talk about the possibility of space colonies. That year, Dan Rather also profiled O'Neill on *60 Minutes*, offering up the professor as the father of the next Space Age.

"Now some serious scientists are talking about whole colonies in space," Rather said during his opening. "Not on the moon, Mars or Jupiter, but on man-made planets. And populated not by scientists and astronauts alone, but by hundreds of thousands of just plain folks looking to get away from an overcrowded Earth running short of energy, water and clean air.

"Far-fetched? That's what we said twenty years ago about walking on the moon. Today, nothing seems far-fetched."

O'Neill offered a special kind of hope at a time of despair over limited resources. His ideas were fantastic, hard to believe and easy to deride. Colonies in space were outlandish, ridiculous. But O'Neill showed they could be real. He had done the math. He had drawn the designs. He even made them part of his curriculum.

On campus, O'Neill was a popular professor, who was warm and welcoming with students, if eccentric, with severely cut bangs and a thin, angular face that made him look a little like *Star Trek*'s Spock, one of Bezos's favorite characters. O'Neill strove to make his introductory course, Physics 112, applicable "to contemporary (your lifetime) problems," he wrote in the notes for his first-day-of-classes lecture. "Not historical. Emphasis on physics relevant to present-day civilization."

For O'Neill, there was no greater question than how to move civilization into space. He had focused his career on this challenge, and this was the problem he wanted his students to wrestle with. And so his exams were peppered with questions asking them to calculate the escape velocity for Phobos, a moon of Mars, how to turn asteroids into habitats, and the energy requirements for space colonies:

"Assume that a small colony of 5,000 people is located in the asteroid belt, 2.7 times as far from the Sun as is the Earth. What must be the diameter of the parabolic mirror used by the colony to bring its land area of 3×10^5 m^2 up to the same sunlight intensity that the Earth receives on a clear day?"

Even though Bezos arrived at Princeton in the fall of 1982 wanting to major in physics, he never took O'Neill's introductory class. Bezos was on the advanced track right from the beginning. He switched to computer science and electrical engineering after he got to quantum mechanics and realized that he was "never going to be a great physicist," he said. "There were three or four people in the class whose brains were so clearly wired differently to process these highly abstract concepts."

But at Princeton, his interest in space was as strong as ever, and he was a regular at O'Neill's seminars, which were open to anyone on campus. O'Neill would "encourage those very capable students who weren't excited by ordinary coursework, inviting them to extra seminars that looked at applying physics to large-scale projects for the benefit of humanity," O'Neill's friend Morris Hornik recalled.

At those seminars, O'Neill posed a pointed question to the students: "Is the surface of a planet really the right place for an expanding technological civilization?"

After Apollo, many thought Mars should be the next destination, and that humans should tick off visiting the planets of the solar system like racking up states on a cross-country road trip. But O'Neill rejected this idea.

"We are so used to living on a planetary surface that it is a wrench for us to even consider continuing our normal human activities in another location," he wrote in *The High Frontier*.

The key question was "whether the best site for a growing advancing industrial society is Earth, the Moon, Mars, some other planet, or somewhere else entirely. Surprisingly, the answer will be inescapable: the best site is 'somewhere else entirely.'"

BY THE TIME Bezos was a senior, he became the president of the Princeton chapter of a student organization called Students for the Exploration and Development of Space. SEDS, as it was known, was started a few years before at MIT by Peter Diamandis, who wanted to increase awareness of space—and would eventually go on to found the Ansari X Prize, a 2004 contest between private companies to launch the first-ever commercial vehicle into space.

At Princeton, SEDS was a small and somewhat lonely group. Despite the popularity of *Star Wars*, which had come out a few years before, space was not high on anyone's list. So the kinds of people it attracted were die-hard space geeks, who did not always fit into the rigid social hierarchy of one of the nation's most exclusive schools.

Karl Stapelfeldt, who became the chapter's president, was two years ahead of Bezos at Princeton and remembered him as being an "interested, loyal SEDS member." The group would meet once or twice a month, raise money for field trips to museums.

"We'd get together and watch shuttle launches, all gathered around the TV," Stapelfeldt recalled. "I always liked to say it was kind of a NASA ROTC. We were all interested in being involved in the space program in some way."

Stapelfeldt did. After receiving a doctorate at Caltech, he eventually became the chief scientist at NASA's Exoplanet Exploration Program.

SEDS, and its relentlessly forward-looking, positive message, was just the sort of thing that interested Kevin Polk, who was a year behind Bezos at Princeton. An inquisitive and thoughtful student, Polk was a devotee of O'Neill and science fiction author Robert Heinlein.

"It seemed to me we had an infinite future in space," he said.

When Polk showed up at a SEDS meeting in the spring of 1985, Bezos had become the group's president and could tell how enthusiastic Polk was about space.

"Jeff and two other guys just sort of looked at one another and said, "Great! You're vice president."

Polk wanted to prove himself by fueling interest in the group and attracting a large crowd to the group's meetings. He had a friend, an accomplished illustrator, design posters advertising SEDS's first meeting of the school year. The posters depicted the tower of Nassau Hall, the university's venerable administration building, blasting off into space with the Princeton Tiger mascot waving in the foreground.

For a student-run group, the fliers were "lavishly rendered," Polk recalled, and Bezos was touched by it—and the note from the illustrator saying she hoped it would suit their needs.

"His jaw dropped and he said, 'What a service-oriented individual your friend is,'" Polk recalled.

The posters worked. More than thirty people came to the meeting. Bezos was pleased to be holding court in the front of the room, talking excitedly about the mission of SEDS and O'Neillian notions of spanning out into space by the millions. Years of reading science fiction came pouring out in a soliloquy that was even further out there than his high school graduation speech.

One way to colonize space, he said, was to transform asteroids into habitats. Yes, he said, people could hollow out the giant rocks, and then live in them. All you had to do, he explained, was use solar mirrors to melt and soften the asteroid, then once it had turned to lava, you injected a massive tungsten tube into the center and flood it with water.

Hitting the molten core, the water would immediately turn to steam, inflating the asteroid like a balloon—and, voilà, there was your habitat. The idea had been around for some time, since futurist Dandridge Cole wrote about turning asteroids into habitats in the 1950s. But as Bezos carried on, a student in the back of the class interrupted Bezos, jumping to her feet with anger.

"How dare you rape the universe!" she shouted. She then stormed out. All eyes turned to Bezos, who didn't miss a beat.

"Did I hear her right?" he said. "Did she really just defend the inalienable rights of barren rocks?"

O'NEILL DIED IN 1992, and never saw his vision get close to becoming reality. But he had touched off a movement by offering hope, his friend Morris Hornik recalled during his memorial service:

"That vast, almost Earth-like colonies could be constructed out of the materials and energy always available in space. That these could become self-sufficient, and that, in his words, 'The human race stands now on the threshold of a new frontier, whose richness surpasses a thousand fold that of the western world 500 years ago.'"

By then, Bezos had left Princeton, moved to New York, where he worked in finance. He eventually took a job at D. E. Shaw & Co., a Manhattan-based hedge fund. Being mired in the cutthroat world of Wall Street didn't leave a lot of time to ruminate about space or to carry on his O'Neillian dreams.

But in 1993, when he was twenty-nine, he went to an auction at Sotheby's that was selling artifacts from Russia's space program. Amazon didn't yet exist, and Bezos couldn't keep up with the deep-pocketed collectors that Sotheby's attracted. Still, he had his eye on a chess set designed to be played in zero gravity. The set, which the catalog described as "a specially-designed mechanical (non-magnetic) chess-set for use in spaceflight," had flown on Russian missions in 1968 and 1969. Sotheby's expected it to sell for between $1,500 and $2,000.

It was a relatively low-cost item in a catalog that featured the first eating utensils used in space, which sold for $6,900, a trio of moon rocks that went for $442,500, and a space capsule that brought in $1.7 million.

Bezos bid on the chess set, but lost out to an anonymous buyer who was vacuuming up many of the items. Still, there was another item that caught Bezos's eye—a hammer that, according to the catalog, was designed "for no rebound of the striking part of the hammer after the stroke, which is extremely important for its use under weightless conditions."

It "was a really cool object," Bezos said later, "because they hollowed out and put metal filings inside the head of the hammer so that when you strike something it doesn't recoil as much."

But he lost out on that, too. Bezos simply didn't have the money to keep pace with more moneyed bidders. Space, and even its artifacts, seemed as far away and inaccessible as ever.

AFTER STUDYING THE staggering growth of the Internet, Bezos left New York for Seattle in 1994 to start Amazon. The company's success was like hitting the lottery; at least that's how Bezos described it. And with his blitzkrieg through the ranks of the *Fortune* billionaire's list, he was freed to pursue almost anything. To those who knew him well it came as no surprise that what he wanted to do more than anything was to start a space company.

Not that he talked about it much. Even his high school friend Joshua Weinstein knew nothing about Blue Origin until reading about it in the news in 2004. Which was odd. Because he had just spent an afternoon in Washington with his old buddy Bezos at the National Air and Space Museum, who said not a word about his ambitions in space.

By chance, Bezos and Weinstein just happened to both be visiting the nation's capital at the same time. Growing up outside of Miami, they'd lived a block apart. "I grew up in his house, and he in mine," Weinstein said. But now they lived on separate coasts—Bezos in Seattle, where he ran Amazon; Weinstein in Maine, where he was a reporter at the *Portland Press-Herald*.

Given his lifelong passion for space, it was natural that Bezos played the role of docent at the museum that day.

"He already knew everything about everything," Weinstein said.

Weinstein kept waiting for people to recognize his companion. He was rich and famous, and had been *Time* magazine's person of the year five years before. But amazingly, no one seemed to recognize him. Or if people did, they left him alone, allowing him to wander around like any other tourist.

Years later, as his fame and fortune grew, Bezos would be followed around by security, men in suits with squiggly wires in their ears, part of the $1.6 million Amazon spent annually to keep him safe. But now he blended in with the midweek crowd, quiet and unassuming, comfortably anonymous.

There were the massive F-1 engines on display that powered the Saturn V rocket to the moon. The lunar roving vehicle. There was also a series of Russian space paraphernalia, donated by the anonymous buyer who had outbid Bezos a decade before on the chess set.

Now a few of the artifacts were at the museum, and the donor had revealed himself: former presidential candidate H. Ross Perot. "He didn't let me win anything," Bezos said years later. Years later, Sotheby's came into possession of another recoilless hammer. This time, the auction house gave it to Bezos as a present.

As THEY TOURED the museum, Bezos didn't mention his unsuccessful foray into space antiquities. Nor did he mention that he had started a space company of his own. He kept it under wraps, as secretive as when he purchased all that property in West Texas.

Blue Origin's website at the time was low-key, revealing little. Bezos's name didn't appear anywhere on it, though it did give away that the company's goal was creating an O'Neillian "enduring human presence in space."

By mid-2004, the company had more than doubled the size of its design team, hiring some of the country's best aerospace engineers from the space shuttle program, Kistler, and the DC-X program, the government's attempt to build a rocket that could take off and land.

"If you have a genuine passion for space and are excited by the prospect of building space hardware, we'd like to hear from you," the site said.

But its "Jobs" page ad was less welcoming, even arrogant. Applicants needed to be "highly qualified and dedicated individuals who meet the following criteria:

"You must have a genuine passion for space. Without passion, you will find what we're trying to do too difficult. There are much easier jobs.

"You must want to work in a small company. If you can happily work at a large aerospace company, you're probably not the right person.

"Our hiring bar is unabashedly extreme. We insist on keeping our team size small (measured in the dozens), which means each person occupying a spot must be among the most technically gifted in his or her field.

"We are building real hardware—not PowerPoint presentations. This must excite you. You must be a builder."

For years, Bezos had been limited to being merely a dreamer lost in science fiction books, O'Neill's teachings, his grandfather's stories. But now he had decided to see what he could do about making those fantasies reality. About one day a week, he was stealing away from his day job at Amazon to quietly indulge his other passion—Blue Origin, where his team was quietly pursuing the hard work of building a transportation network to the stars, creating the heavy-lifting infrastructure that would open the cosmos the way the railroads helped open the American West.

Except for a few flights a year, human spaceflight was as difficult as it had ever been. In the course of Bezos's life, it had advanced very little, if at all. His goal with Blue Origin, then, was to create the infrastructure that finally would allow for humanity to stretch out to the stars.

Once that was in place, "then we get to see Gerard O'Neill's ideas start to come to life, and many of the other ideas from science fiction," he said at a speech years later. "The dreamers come first. It's always the science-fiction guys: They think of everything first, and then the builders come along and they make it happen.

"But it takes time."

He was patient, willing to take his time. "You have to be very long-term oriented," he told Charlie Rose. "People who complained that we invested in Amazon for seven years would be horrified by Blue Origin."

At Amazon, Bezos had been obsessive for years about maintaining its startup culture, even as the company grew, reminding employees that it would always be Day 1 there. In a 1997 letter to shareholders, he wrote that it was "Day 1 for the Internet, and, if we execute well, for Amazon .com." Twenty years later, "Day 1," the name of the Amazon headquarters building, was still a rallying cry. "Day 2 is stasis," he wrote in 2017. "Followed by irrelevance. Followed by excruciating, painful decline. Followed by death. And *that* is why it is *always* Day 1."

On June 12, 2004, he wrote a "Day 1" letter for Blue Origin—"blue" for "the pale blue dot" that is Earth, "origin" for where humanity began—a vision statement outlining the principles that would guide the company:

"We are a small team committed to seeding an enduring human presence in space," he wrote. "Blue will pursue this long-term objective patiently, step by step. By dividing our work into small but meaningful increments, we hope to generate as many useful intermediate results as possible. Each step, even our first and simplest, will be challenging. And each step will lay the technical and organizational foundation for the next."

The first suborbital vehicle would be called New Shepard, he wrote, after Alan Shepard, the first American to reach space. But even then, Bezos had larger ambitions. "At some point, Blue will shift its focus from New Shepard to a crewed orbital vehicle program. Orbital vehicles are significantly more complex than suborbital vehicles, and the transition to orbital systems will stretch Blue's organization and capabilities."

Given the enormity of the challenge, "we believe *local hill climbing* is our best way forward."

That would require a steady approach, into unknown terrain. "We have been dropped off on an unexplored mountain, without maps, and the visibility is poor," he wrote. "Every once in a while, the weather clears up enough for us to glimpse the peak, but the intervening terrain remains largely obscured."

But there would be some bedrock principles to guide them. "Don't start and stop—keep climbing at a steady pace. Be the tortoise and not the hare. Keep expenditures at sustainable levels. Assume spending will be flat to monotonically increasing. Do not fall for the unreasonable hope that the path will get easier as we go up."

Bezos was both a dreamer and a builder, and had created Blue Origin as a laboratory where the two could meld together. In 2005, an interviewer from *Time* magazine asked what he was reading.

He replied that he'd just finished an Alastair Reynolds science fiction novel "about Earth being destroyed by nanobots." That was the dreamer answering. The builder in him was focused on something else: "I've been reading about rocket-engine development."

BLUE ORIGIN'S FIRST test vehicle was an odd contraption, a sort of science fair experiment gone wild. Named Charon, after the Pluto moon, it consisted of four Rolls-Royce Viper Mk. 301 jet engines that the company had acquired from the South African Air Force.

"They were ancient," Bezos said. "I think they were literally 1960s engines. I remember when they arrived at Blue, and the team opened the crates that they were in. Huge spiders came out. Huge South African spiders. And they were like, 'AHHHH!'"

Charon looked like a massive drone and stood on four legs each equipped with a saucer-shaped disk at the foot to help it touch down softly on landing. The engines were pointed down, not sideways, to provide vertical lift—and landing.

On March 5, 2005, at Moses Lake, nearly a three-hour drive east of Seattle, Charon flew. Not particularly high, just 316 feet, or just over half the height of the Seattle Space Needle.

The launch wasn't the main goal. It was the landing Blue Origin was trying to perfect. The vehicle was fully autonomous, meaning it had been preprogrammed with software that allowed it to fly on its own. After hovering at a few hundred feet, Charon descended back to the ground, where it touched down softly, while kicking up a cloud of dust.

It was a small first step. But for the first time, Blue had left Earth—and returned.

5

"SpaceShipOne, GovernmentZero"

URT RUTAN CHOSE the day carefully—December 17, 2003,
the hundredth anniversary of the Wright Brothers' first flight—
to send a signal about the importance of what he planned to
accomplish. The same day that Elon Musk was parading his Falcon 1
rocket down Independence Avenue in Washington, DC, Rutan was getting ready for the first powered flight of the spaceplane he had been building in secret.

He had a cadre of three test pilots working for him to choose from. All
with different backgrounds and experiences. All eager to fly. All fiercely
competitive in what had become an extraordinary race to become the
first commercial mission to reach space.

There was Brian Binnie, the former navy fighter pilot who had experience conducting combat missions over Iraq during the Persian Gulf War
in the early 1990s and held degrees from two Ivy League universities. He
had a runner's trim build and a calm, soft-spoken nature that endured
even when things got hairy in the sky.

Mike Melvill was in many ways Binnie's opposite. At sixty-three, he
was eyeing retirement. A native of South Africa, he had dropped out of
high school and had largely taught himself how to fly. But he was one of
Rutan's very first employees. They had known each other for decades.

Melvill was a natural in the air. The man's instincts in the cockpit were so incredible that Rutan trusted him completely.

Then there was Peter Siebold, the young Generation Xer, who had a round, innocent face that made him look a little like an adult version of Beaver Cleaver, the chipmunk-toothed boy who starred in the 1950s-era sitcom. But he was ambitious and supersmart, and combined his aerospace experience with his engineering background to develop the simulator they were using for Rutan's latest invention, a spaceplane called SpaceShipOne.

"You could not get any more different," Rutan recalled years later. "I wanted all three of these guys to be astronauts."

The curious-looking vehicle was Rutan's entrant in the contest known as the Ansari X Prize, which was modeled after the $25,000 Orteig Prize that Charles Lindbergh won for his epic, across-the-Atlantic flight in 1927. Instead of crossing an ocean, the finish line of the X Prize would be reaching an altitude of 100 kilometers (62 miles), the barrier considered the edge of space.

The winner of the $10 million contest would have to fly a manned spacecraft to that height, land it safely, and then do it again within two weeks. Another rule was that the spacecraft had to be built with private funds—not government money.

The organizers of the X Prize hoped that just as Lindbergh's flight had touched off a revolution in commercial aviation, their contest would spark a new commercial space movement, one that finally would end the government's monopoly on space.

Rutan's spaceship design was, of course, unconventional. All of his planes were. A blunt eccentric with Elvis-like sideburns, Rutan had founded his curious little company, Scaled Composites, in 1982 in Mojave, where his experimental designs often had multiple wings, which sometimes went out and then curved up, making a U shape. Sometimes they had not one fuselage, but three. It was as if his inspiration came not just from the laws of aerodynamics but from Picasso. Rutan had assembled a team of some of the most innovative airplane engineers, who were designing, testing, and then flying the planes they had built, usually within a year.

Instead of launching vertically from a launchpad, SpaceShipOne would be tethered to the belly of a mothership that would fly to nearly 50,000 feet. Once aloft, the mothership, known as WhiteKnightOne, would drop the spacecraft, which would send it plummeting like a baby bird taking what appears to be a suicide dive from its mother's nest. The free fall would last just a few seconds, until the pilot ignited the engines and the spacecraft took off.

The concept, known as air launch, had been around for years, a technique used mostly by the military. Perhaps most famously, Chuck Yeager's Bell X-1 was air launched from a Boeing B-29 before he became the first person to break the sound barrier in 1947 over the same Mojave Desert where SpaceShipOne would fly.

But unlike other air-launched vehicles, Rutan had a special design for SpaceShipOne, an idea that had come to him in the middle of the night. Essentially, the spaceplane's wings would be able to detach from the body of the plane and fold upward in what he called a "feather" maneuver. The upright wings would act like the feathers of a badminton shuttlecock, centering the plane by creating drag for a reentry into Earth's atmosphere so soft that it eliminated the need for a heat shield. Once SpaceShipOne was safely back in the dense air of the atmosphere, the wings would fold back down, and the aircraft would glide back to the ground.

It was a brilliant, revolutionary design that could make the fall back to Earth safer. But if the feather was unlocked at the wrong time, when the spaceplane was screaming upward, for example, it could have devastating consequences.

RUTAN HAD FACED a difficult decision in picking the pilot for the first powered flight. Up to now, the pilots had been flying SpaceShipOne like a glider, floating back to the ground. But on this flight, the plan was to not only light the engine for the first time, but break the sound barrier in what would be one of the most significant tests of Rutan's vehicle.

Rutan was like a baseball manager deciding who should pitch on opening day. Melvill was a trusted friend and accomplished pilot. Siebold had the smarts. But after weighing his options, Rutan went with Binnie. How could you go wrong with a war veteran, who had landed his F/A-18 Hornet on aircraft carriers?

The day of the flight was a beautiful morning in Mojave. The air was still and crisp. If Binnie was nervous, he didn't show it, even though he knew that this was an audition of sorts. If he flew well, then perhaps he'd get the chance to become the first commercial pilot to fly to space.

Binnie, trim and tall in his flight suit, looked as if he was ready for *Top Gun*. As he climbed aboard the spaceplane, he sat patiently while WhiteKnightOne escorted him to altitude. Then, when it was time, he calmly told mission control, "Go for release."

After SpaceShipOne dropped, he lit the ignition and was off. The motor blast pinned him back into his seat, and the engine burned for just fifteen white-knuckled seconds. Still, it was "quite the insult to your senses," he said. "It's a cascade of noise and vibrations. The ship almost immediately complains. You open the gate and you're on this bucking bronco."

Those fifteen seconds were enough. The flight was a success, and Binnie handled the violent force of going Mach 1.2 like a pro, producing a sonic boom that signaled mission accomplished: SpaceShipOne had broken the sound barrier.

"That was a pretty wild ride, Mr. Rutan," he told the ground crew below, as he prepared to come back to Earth.

But as he approached the runway, Binnie was having difficulty keeping the spacecraft level. He was coming in low. Finally he slammed into the ground—hard.

The landing gear splayed outward, like a gymnast doing a split. The bottom of the plane hit the ground like a belly flop, and it tipped over so that the left wing dragged along the tarmac. After skidding a few hundred feet, SpaceShipOne careened off the runway into the brown desert dirt, kicking up an ignominious dust plume.

In the mission control room, Rutan jumped out of his seat and then bolted out to the runway. First responders rushed to the crash site.

Binnie wasn't hurt. But he was furious.

"Damn it," Binnie said, again. And then again. He ripped off his oxygen mask, and went to smack the ceiling of the cockpit before controlling himself.

Rutan was there within moments, trying to calm down his pilot, who was now standing, embarrassed, next to the aircraft he had just crashed.

"Hey, other than that, how was the flight?" Rutan said, trying to soften the blow with humor.

But the former navy fighter pilot was inconsolable.

"Words cannot describe how disappointed—" he began.

Rutan wouldn't hear of it.

"You did a super job," he said. "All we've got there is real minor stuff. It's not a big deal."

The crash was a setback, and, for Binnie, humiliating. This was the first big milestone for the SpaceShipOne program. But now Binnie wondered whether he had just blown his chance to get to space.

The engineers at Scaled would later determine it wasn't Binnie's fault; the flight controls just got stuck on the reentry, overwhelmed with friction in a condition pilots call "stickiction." This was, after all, a test flight, emphasis on "test." The whole point was to push the envelope to see what sorts of problems emerged.

But the competition didn't see it that way, and wasn't afraid to say so publicly.

"He flat didn't fly the airplane," Melvill told *Popular Science* magazine, comments that infuriated Binnie. "He just flew it straight into the ground, like what you would do when flying an F-18 onto the deck."

(Later, in a letter to the editor, Melvill said that he was "deeply hurt by your unfortunate decision to include a comment I don't recall ever making when being interviewed for what I understood to be a completely different article. Brian Binnie is a close friend and one of the best pilots I know." He also wrote that the magazine "used a comment out of context, simply to try to sensationalize a story that was already sensational.")

Despite the crash landing, Rutan couldn't help but be happy—the flight had been a success. They'd broken the sound barrier. They had all these data now on the spacecraft's performance that would help them get to space. That's what he was focused on.

"How was the boost?" Rutan asked Binnie after the flight.

The pilot hesitated a moment. "Uhh," he said. "Pretty wild. The kick and trying to keep the wings level, all of that was pretty dynamic. Just when you think you've got it under control, something different would happen."

In other words, to fly SpaceShipOne, you didn't just need to be a pilot. It might also help to have some experience in the rodeo.

RUTAN WAS THE public face of the program, the brash engineer who said he wanted "to go high because that's where the view is." But until he unveiled SpaceShipOne several months before Binnie's flight, he had

treated it like a classified program, demanding the highest level of secrecy. That was in part because he didn't want word of what he was trying
to accomplish to get out. But it was also because his newest customer—
Paul Allen, the cofounder of Microsoft with his childhood friend Bill
Gates—was a mysterious and reclusive figure with enough wealth to buy
a cloak of anonymity.

Like Bezos and Musk, Allen was an avid science fiction reader as a kid,
fascinated by space. His father was the associate director of the University of Washington's library, where Allen would spend hours after school.
"My dad was just letting me loose in the stacks," he said, sitting in a conference room outside his Seattle office, with a view of the Space Needle.
"I loved it." He read Willy Ley and books about Wernher von Braun's
V-2 rocket. He became fascinated with engines, and turbo pumps and
propellants.

Allen knew all the names of the Mercury 7 astronauts as if they were
the players of his favorite baseball team, and he wanted to be an astronaut
when he grew up. But then in the sixth grade, he no longer could see the
blackboard, even from his front row seat. His nearsightedness meant "my
dreams of being an astronaut were over," he said. "Somehow I knew you
had to have perfect eyesight to be a test pilot, and so that was it for my
astronaut career."

He once tried to launch the arm of an aluminum chair by packing it
with powdered zinc and sulfur and firing it from a coffeepot, he recalled
in his memoir, *Idea Man*. It didn't work.

"Turns out the melting point of aluminum was lower than I understood," he said.

As an adult, his passion for space continued. In 1981, he went to the
Kennedy Space Center to watch the first shuttle launch. "The sound was
unbelievable," he recalled. "The air was vibrating, and you could feel
compression waves going into your chest. . . . You could feel the heat
from the engines on your face." Allen watched it alongside the tens of
thousands who had packed the Florida coastline, so many yelling, "Go!
Go! Go! It was so inspiring."

After he had cofounded Microsoft, Allen was one of the richest men
in the world, free to pursue his passions. An avid sports fan, he bought
the Portland Trailblazers and the Seattle Seahawks. In Seattle, he opened
the Museum of Pop Culture. He was also keenly interested in aviation

and amassed a collection of historic World War II fighter planes that would eventually go on display at his Flying Heritage & Combat Armor Museum.

In 1996, the year the X Prize was announced, he went to Mojave to visit Rutan and chat about his plans to build a supersonic jet that could fly over the atmosphere. They stayed in touch, and two years later, Rutan flew to see Allen in Seattle to propose something even more ambitious—his plan to develop SpaceShipOne. It would be another two years before he felt he had a design that could work.

Allen was sold and would invest more than $20 million in a venture that, if successful, would pay out half of that.

Rutan knew that if word of his latest project leaked, he'd inevitably be laughed at, ridiculed. And he didn't want the distraction of anyone, not his fellow aviation engineers, not the press, not anyone, telling him what he was trying to do was impossible.

Then again, he had a different relationship with the word than most. One of Rutan's favorite sayings was that it's not research unless half the people involved think what you're trying to do is impossible. He urged his engineers to take risks and told them that "a true creative researcher has to have confidence in nonsense."

They—the doubters, the skeptics—said he couldn't build Voyager, which in 1986 became the first airplane to circle the globe without stopping in a trip that took 9 days, 44 minutes, and 30 seconds. And that's what they would say now.

Rutan had become one of the most accomplished aerospace engineers of his generation—with several of his aircraft retiring to the National Air and Space Museum. But soon he was looking for the next frontier. Upset with what he saw as the retreat of America's space program, he would tell the *New York Times* that "NASA has almost ground itself to a full stop." To Rutan's thinking, the agency had become another bloated government bureaucracy, subject to the fickle whims of Congress and ever-changing administrations.

The space shuttle, which was supposed to fly safely and affordably, had accomplished neither, and was viewed by its skeptics as an expensive death trap that had killed fourteen astronauts in two catastrophic explosions. Worse, it had sent NASA scurrying into retreat, scaring the once bold agency into a risk-averse bureaucracy.

As far as Rutan was concerned, the government had abdicated its monopoly in space. Only the private sector could advance spaceflight now,
he thought. It could innovate and move quickly in a way no government
agency could.

So, he would build the world's first commercial spaceship. That was
the secret, the project he was keeping covert, shielded from the derision
he knew it would attract. Scaled Composites, the cynics would say, a company of just a few dozen people, could not start a manned space program.

Until it did.

ON JUNE 21, 2004, six months after Binnie broke the sound barrier,
and several test flights later, Rutan was ready to try a test flight to space.

After the hard landing, Binnie didn't think his chances of being picked
were very good. Even though it wasn't his fault, he was beginning to feel
"benched on the sidelines," as if the prevailing attitude at the company
was that he "crashed it because I flew it like a damn navy guy into the
ground. The entire undercurrent was clearly I don't have the right stuff:
'Just look at the mess he made of the vehicle.' That stigma and that undercurrent was very much the persistent attitude of the company."

The relationship between the test pilots had grown strained as they
competed for the available slots. "Instead of us working together, passing
on lessons learned and because of the secrecy it pitted us all against each
other," Binnie said. "It was a very corrosive environment in that regard."

On the day the announcement was made, Siebold was in Binnie's office, discussing an issue with the aircraft's avionics, when Binnie received
an e-mail from the flight test director. Binnie knew this was it—the announcement he'd been dreading.

"Do I want to read the bad news now or wait until after lunch?" he
thought to himself.

He opened it.

"Look at this," he said to Siebold, trying to stay casual. "Mike's the
next guy up."

Siebold, normally able to retain a test pilot's preternatural calm no
matter what, "turned beet red, totally agitated," Binnie recalled, "totally
flummoxed that he somehow had lost out on an opportunity that apparently he had lobbied for quite hard."

The team just had a higher comfort level with Melvill, Rutan's trusted confidant for nearly three decades. Of all the test flights, this was the big one—the first attempt to reach the 100-kilometer barrier of space. If anything went wrong, Rutan knew he could rely on the experienced pilot, despite his age. And Melvill had just recently proven, once again, that he had the right stuff to pull off a stunt as crazy, and dangerous, as this.

On an earlier test flight of SpaceShipOne, the flight navigation system went out just as he had hit the ignition switch and was screaming upward, almost perfectly vertical. Everyone in mission control figured Melvill would just cut the engine, end the flight, and come back safely. Flying that fast without a navigation system would be insane.

Instead, Melvill kept the motor on for the full fifty-five-second burn, flying 3,400 feet per second, or faster than a speeding bullet, all while flying essentially blind. His only navigation tool was to look out the window from the corner of his eye at the horizon. He nailed the flight, and the landing, leaving Rutan, a hard man to impress, in awe.

On the ground, Rutan celebrated with his friend, telling him about what it was like watching him from mission control.

"Everybody's expecting you to abort," Rutan said. "And I said, 'He's going to run it at least thirty seconds.' And then I said, 'No, he's going to run it at least forty seconds.' And then I said, 'No, he's going to run it all the way!'"

"Damn right," Melvill replied.

Rutan acknowledged to *Popular Science* that "in some places, that would get a test pilot fired. In this case, I thought it was a positive that Mike could hang in there and press on."

But to Siebold, the move was evidence of unnecessary risk, not bravery, and he had misgivings about Melvill's selection for the attempt to reach the threshold of space.

This was going to be the big one—if Melvill was successful, he'd go down in history as the first pilot to fly a truly commercial, nongovernment vehicle to space and back. Siebold wrote in an e-mail that Melvill was a "cowboy" who flew loose and risky, according to Julian Guthrie's book on the X Prize, *How to Make a Spaceship*. Rutan got ahold of the e-mail and showed it to Melvill to pump up his competitive juices against the younger rival.

"See what you're up against," Rutan told his pilot, according to the book.

Binnie didn't know about the e-mail, but Siebold tried to convince him to join the chorus against Melvill. "He thought it was reckless and cowboylike behavior, and he tried to get me to sign up to that," Binnie recalled.

He refused. Stuff happens in the air, especially in experimental aircraft, and Binnie said he would have done the same thing in Melvill's situation.

If Binnie had been cast as the guy who crashed the plane, Siebold had been branded as too cautious. On one of his earlier test flights, he faced a dilemma. After SpaceShipOne was released, he noticed that one of the wing flaps appeared to be stalling. If he flew, he feared he wouldn't be able to control the spacecraft. But if he didn't, he'd land with a tank full of fuel that made the aircraft too heavy for a safe landing.

As he talked it over with mission control, crucial seconds were ticking away—as he was falling faster and faster. Finally, mission control told him he needed to light the engine. Landing with that much fuel was just too dangerous. Siebold did, and flew safely.

When Siebold was back on the ground, Rutan greeted him warmly and congratulated him. But because Siebold had waited so long to light the motor, he didn't reach the altitude he was supposed to, meaning he didn't achieve the goal Rutan had set for him. The deliberative, careful approach was perhaps the right way to handle a potentially serious problem. No one wanted a dead pilot. But it was also the opposite of how Melvill had just gone for it when he was faced with a problem.

Still, for this first flight to space, one of the members of the team advocated for Siebold, whose flights in the simulator were impressive.

"Yeah," Rutan concurred. But he still had this reservation about Siebold: "He might quit."

"Pete didn't achieve the goals of his first rocket-powered flight in SpaceShipOne because he couldn't bring himself to throw the switch and light the motor at the right time," Rutan said later. "Mike and Brian had come off the hooks and thrown the switch."

Melvill would be the pilot for the first launch to space. "A gutsy call," Allen recalled in his memoir. "Despite Mike's 6,400 hours of flight time, this would be well beyond anything he had done."

The decision left Siebold disappointed.

"I think every one of us wishes that we could be on that vehicle, and fly that really challenging flight," Siebold told a documentary film crew from the Discovery Network. "This flight is what's going to get the attention of the world. This is the flight which says, 'Hey, NASA, we're here.'"

At a press conference the day before the flight, the three pilots stood shoulder to shoulder in their flight suits, presenting a unified front. Rutan announced his lineup: Binnie would fly WhiteKnightOne, the mothership; Melvill would fly SpaceShipOne, with Siebold as his backup.

Rutan acknowledged the danger of what they were trying to do, saying, "We are willing to seek breakthroughs by taking risks. And if the business-as-usual space developers continue their decades-long pace, they will be gazing from the slow lane as we speed into the new space age."

Taking the podium, Paul Allen said they were chasing history.

"Tomorrow, we will attempt to add a new page to the aviation history books. If our attempt is successful, SpaceShipOne's pilot will become the first civilian pilot to ever cross the boundary of space in a completely privately funded vehicle."

Left unsaid was the fact that he was nervous about Melvill's safety. So was his wife, Sally, a pilot herself, who just before the flight implored her husband just to "come home to me."

"I've had any number of people, guys as well as women, come to me and say, 'Well, how can you let him do that?'" she said in the Discovery Channel documentary *Black Sky*. "I don't believe I have the right to tell him what he can and can't do. Even if I think it is high risk and life threatening and whatever. I mean, this is his ultimate joy."

Melvill knew the risks—and how terrified his wife was. He'd been a test pilot for years, but he noticed how she'd be a lot more nervous the older he got. This was unlike anything he had ever done. The spacecraft would be flying three times the speed of sound, faster than it had ever flown before to a height of 62 miles. And the crew at Scaled Composites had made some last-minute adjustments to the vehicle that they hadn't yet had a chance to test.

Sitting in the cockpit just before the flight, Rutan came over for one last pep talk.

"This is the big one, Burt," Melvill told him, as the men exchanged an extended handshake. "Thanks so much for the opportunity."

"We've got the right guy," Rutan replied. "It's just an airplane. Don't worry about it."

Along the tarmac, thousands of onlookers had gathered, many arriving in the predawn darkness to witness what they knew could be historic—or disastrous.

In the air, Melvill seemed relaxed and ready. When he hit the ignition, Sally, watching through a pair of binoculars, yelled out, "Go, Michael! Go, babe!"

The flight started with the usual violent jump, as Melvill fought to point SpaceShipOne straight up. But just eight seconds into the flight, he got pushed off course by the wind. As he struggled with the controls, the engine roared angrily, shaking the spacecraft. Then, he heard a series of bangs that set his imagination running wild. Had a piece of SpaceShipOne broken off?

Still, he kept climbing and climbing until the engine shut off and he floated on. His early problem had knocked him more than 20 miles off course. But it appeared that he had—just barely—crossed the 62-mile threshold.

"Wow," he told mission control. "You would not believe the view. Holy mackerel."

Rutan turned around to congratulate Allen, shaking his hand, flashing a wide-eyed smile. But soon they realized there was another problem. A trim flap on Rutan's feather system, the shuttlecock-like device that was to deliver him to ground safely, was malfunctioning. If the stabilizer didn't work, SpaceShipOne would enter a violent spin, and Melvill could easily be killed on the reentry.

This was the moment he was supposed to be celebrating. Melvill had made it to space. Outside he could see the thin layer of the atmosphere and the curvature of Earth. He could see the deep, vast blackness of space. But instead of taking in the moment, he worried about how he'd get back to Sally.

On the ground, she was a jumble of nerves, now huddled around a walkie-talkie, her hands clutched together as if in prayer, listening to her husband and mission control sort out the mess he was in.

"This is not good," someone in mission control said.

Melvill tried to adjust the stabilizer system again, and after a few seconds it worked. He'd be okay. Relieved, now he could enjoy the little time he had left in space, before gravity pulled him back to Earth. He pulled out a couple handfuls of M&M's that he had secretly stashed in the left shoulder pocket of his flight suit, which in the weightless environment floated through the cockpit, pinging gently off the windows. He finally allowed himself a moment to enjoy a view that only the some four hundred people who had been to space before him had ever seen.

Moments later, as he guided SpaceShipOne to a flawless landing, Sally Melvill, her hands clutched again in front of her, was near tears. "Oh, thank you. Thank you. Thank you!" she said to no one in particular.

After her husband emerged, she broke down in his arms.

"Thanks for coming home," she said, sobbing. "Can we grow old together in rocking chairs?"

He said they could. He was, as of now, retired as a SpaceShipOne test pilot. He had made history, earning the first ever "commercial astronaut" wings from the Federal Aviation Administration (FAA).

Rutan was ecstatic, and said later that he was glad it was Melvill in the cockpit, and not anyone else. "The more experienced people would have aborted two or three of the flights, which would have set us back many months," he said.

Now, he had proven that a small band of dedicated, passionate rocketeers could pull off a feat no one thought possible. But beyond that the flight didn't just symbolize the emergence of the commercial space industry, a New Space movement, but, he felt, the obsolescence of NASA.

After Melvill landed safely, Rutan grabbed a sign from the cheering crowd that summed up exactly what he was thinking and waved it proudly:

"SpaceShipOne, GovernmentZero."

WHILE ALLEN WAS a space enthusiast, he made an uncomfortable pioneer. The SpaceShipOne flights made him realize that he did not have the stomach for the risk of human spaceflight, and instead of enjoying the historic feats, he was petrified that the pilots flying his spacecraft would be killed.

During Binnie's first powered flight, Allen was overcome by "a wave of dread," he wrote. When developing computer software, "your worst outcome is an error message. Now I knew the person whose life hung in the balance, and I found that hard to handle." And when Binnie crashed, he felt as if his "heart was in my throat," wondering whether Binnie was hurt.

Just before the X Prize flights, Allen had received a call from Richard Branson, the billionaire founder of Virgin Records and Virgin Atlantic, who had begun a space venture of his own and was looking for a vehicle to acquire. If Allen was a recluse who valued secrecy and shuddered at the dangers of spaceflight, Branson was his opposite, a thrill-seeking, media-savvy marketer who pursued one adventure after the next.

Branson, who started an airline and a train company, and had been on several daredevil, record-setting hot-air balloon rides, was desperate to start a company that could help push what he saw as the ultimate frontier of space. Smitten with SpaceShipOne, and confident that Rutan could build him an even bigger and better spacecraft, one capable of taking fleets of tourists into the cosmos, Branson made Allen a generous offer for the rights to the technology behind SpaceShipOne.

"Flying test pilots, I understand," Allen recalled. "But paying-man-on-the-street-type passengers, I wanted to leave that to someone else."

Allen would see the X Prize through, but then he was eager to move on to other ventures. So, he sold the rights in a deal worth up to $25 million over fifteen years. Branson, who had added Virgin Galactic to the list of companies he ran under the Virgin brand, quickly had the Virgin logo painted on SpaceShipOne just in time for its X Prize flights.

BY SEPTEMBER, RUTAN's team was done with test flights and was ready to go for the money. To win the $10 million Ansari X Prize, SpaceShip-One would have to fly to space twice in two weeks, while reusing at least 80 percent of the vehicle.

Rutan decided to go with Siebold for the first prize flight. Binnie would be the backup—and was beginning to fear that because of the crash landing they'd never let him fly again. "I understand the concept of 3 strikes you're out," he wrote in an angry e-mail to the flight director. "I just don't know what the count is anymore."

Melvill had done his part. And it wasn't clear that his wife, Sally, could take his flying another harrowing ride like that, no matter how thrilling it was. Siebold had been upset that he wasn't chosen for the first space attempt, and had been training for this mission for more than three years.

But suddenly Siebold was getting second thoughts. His wife had just had a baby, and in the weeks leading up to the flight, he had a brush with a potentially serious illness. But he also felt the plane was unsafe and needed further testing.

There were plenty of signs that the engineers were still working out the kinks of the aircraft. Melvill had gone 20 miles off course on one flight, and had trouble with the stabilizer on the other. The flight controls froze up on Binnie, causing his crash landing. A new father, Siebold had a tough decision to make. As hard as it was to disappoint Rutan and the rest of the team at Scaled Composites, he couldn't go through with it. The flight was just too dangerous.

"Peter to his credit had lost faith in all the haste to light a wick under the rocket motor," Binnie wrote in an unpublished memoir titled "The Magic and Menace of SpaceShipOne." "He felt that it was unsafe, insufficiently tested and poorly understood. To him, that was three strikes in a critical spaceship system and not worth the risk."

Scaled Composites would tell the public that Siebold was merely sick, and never let on about his concerns that the spacecraft was unsafe and not fully tested. It was already mid-September. The first flight was just days away, and the whole point of the exercise, after all, was to convince the public that spaceflight could be made so safe it would be routine.

With days to go before the flight, Rutan had to again ask his trusted and tested friend Mike Melvill. After the last harrowing flight, Rutan knew Melvill had a "feeling of relief that I didn't die today, and that I won't die in the program because I'm done." But then the team "had to go to him to fly another spaceflight."

Binnie was furious and stormed into the flight director's office, demanding to know "when is the backup not the backup."

The flight director "cut to the chase and said that the landing from last year did not sit well with the Boss and suggested his attempts to work me back into the lineup were thwarted," Binnie wrote in his memoir. "So there it was. And it was worse than I had ever imagined. I felt defeated."

While Rutan said that the controls had frozen up on Binnie during the landing, he said "we were not able to go to Brian because Brian had been so tied up with rocket development and everyone questioned his proficiency when he made the hard landing. . . . We couldn't put Brian into the flight because we didn't think he was ready."

Sally Melvill burst into tears when she got the news that her husband would fly again.

"To be honest, I was very irritated," she told the Discovery Channel. "I had settled my mind that he was not going to do another flight. So I had my emotions really where I needed them to be. And to start working and trying to get mentally prepared—and Michael had the same problem."

It wasn't just the mental adjustment that concerned them. He hadn't been preparing physically. To get ready for the violent churns of the flight—the increased gravitational forces that would strain their body—the pilots had trained hard in test airplanes. They put themselves into dizzying tailspins, banked hard, flew upside down—all in an effort to get their body ready.

Melvill hadn't had time to train properly, and superstitious as he was, he wondered whether this flight might be one too many. Shortly after Rutan said he needed him, he told his wife that he "wondered if I was pushing my luck doing a second flight. Am I asking too much?"

Sally Melvill wondered that, too.

THE FLIGHT ON September 29 started as expected. WhiteKnightOne climbed into the early morning Mojave sky. It released SpaceShipOne, and then a few seconds later the rocket engine ignited, pinning Melvill back into his seat as he screamed almost straight up in a picture-perfect beginning.

From the ground, it looked just as it should. "He's straight!" Sally Melvill yelled. "He's straight! He's absolutely dead straight."

But then SpaceShipOne started to roll. It was slow at first, but the higher he climbed, the faster the spacecraft spun, until soon it was whirling uncontrollably. The nose of the craft was still pointed toward the sky, but the wings were whipping around so quickly that the sun was flickering in the cabin, as if someone was turning a light switch on and off.

Melvill kept his head straight, focusing only on the control panel in front of him. He didn't dare look out the window. Seeing the world spin below him would only rattle his nerves—and make him sick. Just as he had on the earlier flight when his navigation system went out, he kept the engine firing. Spinning be damned. He was still climbing toward space.

Finally, he crossed the 62-mile boundary and by firing thrusters on the spacecraft, was able to slow the rate of the spin, just in time for reentry.

Once again, it had been a harrowing ride. But once again, Melvill stayed in the saddle all the way to space.

One flight down for the $10 million prize, one to go.

THE NEXT DAY, a Thursday, the team gathered for a meeting. Everything seemed a go for the second flight, which they'd push ahead for the following Monday. Despite not having flown since the crash, ten months earlier, Binnie tried to remain sharp, spending hours in the simulator, hoping he'd have a chance, even if he felt he was a long shot.

They went through the logistics. The avionics looked good. Everyone was happy with the flight profile. The rocket seemed good. They were all set, and about to wrap up, when the crew chief raised his hand.

"Burt, I need one last piece of information," he said. "And that's the pilot."

After an uncomfortable silence, the test flight director said, "Well, Brian is, of course."

Melvill was done. Siebold had pulled himself out of the program. Binnie was really the only one left. He felt he was the pilot of last resort, as if, as he said, everyone in the meeting was thinking, "We have nowhere else to turn but to the guy who crashes spaceships."

With the flight just days away, he didn't have time to brood. And the former navy pilot wanted to redeem himself.

Now out of the running, Melvill generously helped prepare him, taking him up for test flights in his plane.

On the morning of the flight on October 4, Binnie made his way toward the spacecraft and saw his mother-in-law. Carrying a cup of coffee, she came to give him a good-luck hug. But as she reached her arms around him, she ended up spilling coffee down his back. Binnie didn't have any time to change—and didn't have another flight suit anyway, "so I manned up with this sticky mess all over me."

He was wet, and the smell of sugary coffee overwhelmed the cockpit. But he was ready.

The WhiteKnightOne mothership released SpaceShipOne. Instead of waiting for mission control to give the all clear before igniting the engines, Binnie flipped the switches almost immediately, not wanting to lose too much altitude, and zoomed right by the mothership, where a surprised engineer on board yelped, "Holy crap! That was close!"

But other than that, the flight went as smooth as could be. He reached higher than Melvill had gone in his two earlier flights, setting a new record for a commercial spacecraft.

Now partners, Allen and Branson came to the Mojave Desert to witness Binnie clinch the victory, and could not have made a more different pair. Here was Branson with his gilded flowing hair, Virgin Islands tan, standing next to Allen, pale and pasty in baggy jeans.

"Paul, isn't this better than the best sex you ever had?" Branson asked him, as the spaceship climbed higher.

"If I was this anxious during any kind of interpersonal activity, I couldn't enjoy it very much," Allen thought.

Binnie nailed the landing—no belly flops this time—a nice soft touchdown right in the middle of the runway.

"He greased it on like an air force pilot, not a navy pilot," Rutan said. "He flew the only perfect flight of SpaceShipOne. I was very proud of him."

During the celebration, Rutan once again took aim at NASA.

"I was thinking a little bit about that other space agency, the big guys," he said. "I think they're looking at each other now and saying, 'We're screwed.'"

To underscore the point, NASA wasn't flying at all at the time. Space Shuttle Columbia had disintegrated years before, killing another seven astronauts. The shuttle program was grounded while investigators tried to figure out what went wrong. In all of 2004, the US government did not fly a single spaceflight.

In fact there were only five trips to space that year. The Russians flew two. Burt Rutan flew the other three.

This was the triumph of the little guy, the individual, a uniquely American moment. "I just thank God that I live in a country where this is possible," Binnie said.

For their flights, the FAA had largely stayed out of the way. Since no one but the government had ever even attempted to fly to space, there weren't any laws preventing what they were doing. And what regulations were in place were lax. For now. Congress would surely take notice, and hold hearings to discuss what regulations should govern this new industry.

But all that was for a later day. Now was a time to celebrate. Rutan gathered the Scaled Composites team in front of the hangar.

"The important thing about today's accomplishment is this is not an end," he said, as Allen stood next to him. "It's just a very good beginning."

Rutan and Allen then popped champagne, letting the bubbly flow. Rutan took a big swig directly from the bottle.

At the same time, Branson was already dreaming of the next space-craft, SpaceShipTwo. Rutan was on board to build it. But this time he'd be doing so for Sir Richard, the playboy who always liked a splash. Their new spaceship wouldn't be built so that it could win a prize. It would be designed for luxury, for as many as six passengers and two pilots, with the kind of first-class touches that Branson was known for at Virgin Atlantic, his airline.

At the moment, the spacecraft was still just a vision in his head. But he couldn't wait to show it off.

PART II

IMPROBABLE

6

"Screw It, Let's Do It"

OR HOURS ON end there had been nothing but the perilous Atlantic below, but now they could see the inviting shoreline and the lush and verdant Irish countryside just beyond. They had made it more than 3,000 miles, crossing an ocean in a hot-air balloon.

Some twenty-four hours earlier, Richard Branson, then thirty-six years old, and his pilot, Per Lindstrand, a ballooning expert and aeronautical engineer, had taken off from near Sugarloaf Mountain in Maine, then crossed the ocean for a first-of-its-kind transatlantic flight that would land them in the record books. All they had to do now was land.

It was a foggy day, July 3, 1987. The towering, twenty-two-story-high balloon, with its Virgin logo, emerging from the clouds and now hovering over the most pastoral of settings, made for something of a surreal scene. It only got more bizarre when the wind picked up, swirling in punishing gusts. Instead of touching down softly, the balloon's pressurized capsule crashed in a field next to a quaint country cottage, dragging along the ground so hard that the fuel tanks ripped off. Then, the balloon suddenly lifted off again, just missing the cottage and some nearby power lines.

"With no fuel tanks we were utterly out of control," Branson recalled in his memoir, *Losing My Virginity*.

As they headed back toward the sea, Lindstrand decided to try to land on the beach. But once again the wind had other plans, ushering the balloon back out over the water. They came down with an unsettling splash, as the balloon tilted sideways, acting like a giant sail that dragged the capsule through the water.

Lindstrand pulled the lever that was supposed to release the balloon from the capsule, but nothing happened. He tried again. Still nothing.

"The next thing, we found ourselves being hurtled through the water at something like 100 miles per hour with water coming into the capsule," Branson said at the time. "We climbed out on to the roof, and the capsule started rising, and Per threw himself off at about 60 feet."

As Lindstrand prepared to jump, he yelled at Branson to do the same. Just then, the wind lifted the balloon up, pulling the capsule off the sea surface. Lindstrand jumped.

Branson watched him plunge into the cold water off the Northern Ireland coast, and realized "with horror" that it was too late for him to jump. The balloon was flying upward again. He was alone in a balloon he hardly knew how to fly. Frantic, he tried to call for help, but the radio was dead. He tried to collect himself and figure out how to get out of this mess.

"Standing amid the swirling white cloud, I felt an overwhelming sense of loneliness," Branson wrote.

And dread. He scribbled a quick note to his young family, telling them he loved them. Standing atop the capsule, riding a runaway balloon, all alone, he felt it was quite possible he'd never see them again.

THE IDEA TO cross the Atlantic Ocean in a balloon had been Lindstrand's idea. But Branson had immediately warmed to the adventure. He was young and impulsive, following what would become a mantra for his life and career (and even the title of one of his books): "Screw it, let's do it." But he had a genuine itch for adventure, which had been shaped in large part by his mother, Eve, who always promoted self-reliance in her children.

Eve Branson had an independent streak of her own that stretched back to her childhood, growing up during World War II. She had been training as a dancer, but as the war broke out, she felt she needed to contribute in some way. She heard Britain's Air Training Corps was looking for glide flight instructors, and so she signed up. There were only a couple of sticky problems: she had no idea how to fly a glider, and she was a woman seeking a job that was restricted to men. Undeterred, she disguised herself as a man, showed up, and eventually was allowed to fly.

Her first flight, however, almost ended with a crash. And when she did finally successfully land, "I stepped out of the glider, shaking, and was met by an ashen-faced crowd of officers and cadets who had run across the airfield to greet me," she wrote in her memoir, *Mum's the Word*.

Another of Branson's heroes was Douglas Bader, a famous World War II Royal Air Force pilot who had lost his legs in a crash but continued to fly during the war, eventually commanding a squadron. In 1941, he was forced to parachute out of his Spitfire during combat and was taken prisoner by the Germans. Released after the war in 1945, he was greeted as a hero and eventually was knighted for his service to the disabled.

Bader had become close friends with Branson's aunt, and to Branson, he was like an uncle. As a child, Branson enjoyed stealing Bader's prosthetic limbs when he would take them off to go swimming. The former fighter pilot would chase after him, running on his hands.

"He was a hero of my childhood days," Branson recalled. "He was a larger than life character. . . . He, my aunt and my mother would all go off flying planes and take off in the field near the house and do acrobatics above us. And some of my yearning for flying and adventure came from him."

And also from Captain Robert Falcon Scott, a distant relative, navy officer, and explorer who led an expedition that attempted to become the first to reach the South Pole in 1912. The team completed the journey, only to discover that a Norwegian expedition had beaten them there. Scott and the rest of his crew died while trying to make their way back from Antarctica.

So, later in life, when Lindstrand came to Branson with this crazy idea to fly a balloon across the ocean, it didn't seem so crazy. A transatlantic balloon trip would be just the sort of thing that would have impressed Bader—and his mother, if she weren't so worried. If she could dress up as a man and train pilots without any experience, he could do this.

But it wasn't just the promise of the thrill that attracted him. A stunt like this would generate all sorts of publicity for Branson and his young airline, a brash upstart of a company he had founded on something of a whim three years earlier. Frustrated with how the industry treated its passengers, herding them into cramped planes, with poor service and frequent delays, he thought there had to be a better way to fly.

His exasperation with airlines' cavalier attitude toward their passengers came to a head after a flight from Puerto Rico to the British Virgin Islands was canceled because not enough people were on it.

Desperate to get there where a "beautiful lady" was waiting for him, he chartered a plane and found a chalkboard. "Virgin Airlines" he wrote on top, and then "$39 one way to BVI."

"I went round all the passengers who had been bumped and filled up my first plane," he later explained.

He called Boeing and asked to rent a plane, and that was the beginning of Virgin Atlantic. But now his young airline was struggling to compete against British Airways, a behemoth that was trying to squash Branson's young and plucky upstart. "We needed to come up with fun ways of promoting the airline, getting Virgin on the map," Branson recalled.

A daring balloon ride would do it.

HAVING WATCHED LINDSTRAND jump, Branson was alone in the balloon when it began rising again into the clouds above the water. Not sure what to do, he climbed out on top of the capsule and, with a parachute on his back, thought about jumping. But that seemed rash. He had a giant balloon over him—*that* would be his parachute. He hadn't trained a lot on the balloon, but he spent enough time to have a sense of how to bring it down.

Carefully, he fired the burner, trying to bring the balloon down while peering out at the sea, trying to judge the distance. Just before the capsule hit the water, he inflated his life vest and jumped. Within minutes, a Royal Air Force helicopter was overhead, ready to pull him out of the icy ocean, as the balloon "soared back up through the cloud like a magnificent alien spaceship, vanishing from sight," he wrote. They rescued Lindstrand, too, who had been in the water for two hours before they found him, shivering and nearly frozen.

A close call. But a dramatic ending to an extraordinary journey—one that achieved his goals of setting the record, ginning up loads of publicity while also creating plenty of material for his PR machinery to work with.

As he recalled years later, Virgin Atlantic used the opportunity to take out a full-page newspaper ad that said something along the lines of "Come on, Richard, there are better ways of crossing the Atlantic."

BRANSON HAD NAMED his companies "Virgin" because he and his fellow founders were virgins in business. He was a high school dropout who, dyslexic, couldn't read a spreadsheet. But he was a genius at generating publicity for a long string of companies, starting with the magazine he dropped out of boarding school to run. Then there was the mail-order record company, followed by a record shop, then a recording studio.

In 1977, when Branson was twenty-six, Virgin Records signed the Sex Pistols, a punk band that had just been fired by its previous label for the musicians' raucous behavior. Branson put out the band's album *Never Mind the Bollocks* and plastered the cover in his record shop window. A young police inspector with the Nottingham Constabulary arrested the manager of one of Branson's stores for violating the Victorian-era Indecent Advertising Act over the word *bollocks,* English slang for "testicles."

The charges were dropped. And the incident, covered in big, bold headlines by the British tabloids, generated loads of publicity for the young punk band and its chief promoter. But a scandal over a saucy word only went so far. Nothing seemed to get more attention than a wild, record-setting, adventurous ride that was potentially fatal. Especially if the person risking his life was rich.

In 1985, Branson attempted to set the fastest time across the Atlantic in a boat. He agreed to go on the journey in part because he "relished the chance to promote our new airline. A successful Atlantic crossing would attract publicity in both New York and London, our sole destinations."

The first attempt ended in disaster, after the Virgin Atlantic Challenger roared across the choppy ocean for three days, pounding through the waves with such force that it "was like being strapped to the blade of a vast pneumatic drill." Branson and his team were within just 60 miles of the prize, when they hit a storm and crashed into a wave that broke the boat's hull, forcing the crew to bail from the sinking ship.

The next attempt, however, was successful. The Virgin Atlantic Challenger II broke the record, making the journey in 3 days, 8 hours, and 31 minutes, bringing back the honor to Great Britain, after the United States had held the record for years. To pull off the stunt, Branson had missed the birth of his son, Sam. But he could justify it over the headlines he had created.

To celebrate, Prime Minister Margaret Thatcher joined Branson on the ship, a cross between a cigarette boat and a yacht, and they floated down the Thames, waving to the crowds, gaining the sort of attention no ad or marketing campaign could ever achieve.

He'd carry on that way for years—dressing up as a bride to sell his new bridal shops, getting stranded in the Algerian desert during an attempt to circle the globe in a balloon, crashing in the Canadian Arctic on a successful attempt to cross the Pacific in a balloon that was supposed to land in southern California.

At a celebration of Virgin Atlantic's twenty-first birthday, he slung a scantily clad Pamela Anderson awkwardly over his shoulder so that her breasts fell out of her revealing red dress in what may or may not have been an intentional flashing before the cameras. To launch Virgin Cola, he drove a tank through a wall of Coca-Cola cans in New York's Times Square. And when he finally signed the Rolling Stones to Virgin Records, completing a lifelong dream, he said the most memorable part "was the hangover I had the day after."

CBS's *60 Minutes* dubbed him "the billionaire stuntman." The *New York Times* called him "a one-man publicity circus." It was all in the service of his ever-growing empire of companies that would eventually include a wild, scattershot portfolio—Virgin Mobile, Virgin Money, Virgin Wines, Virgin Trains, Virgin Casino, Virgin Books, Virgin Racing, Virgin Sport, Virgin Media, Virgin Hotels, Virgin Holiday Cruises, Virgin America, Virgin Australia—which, taken together, personified corporate attention deficit disorder as much as ambition.

There was, however, a method to this madness. A common thread that ran through all these seemingly disparate ventures: the embodiment of cool and sexy, and the very Branson "screw-it-let's-do-it" freedom that tiptoed along the fine line separating recklessness from brilliance. So, there would be no Virgin Tax Prep, no Virgin Dentistry, no Virgin Bow Ties. Only that which held quixotic promises of the good life that lay somewhere in between the never-grow-old teenager's innocent idealism and the Sex Pistols' convulsing, rocket-thrust riffs.

But none of it—not the speedboats, not the ballooning, not the Rolling Stones (well, maybe, the Rolling Stones)—could compete with the venture he was now pursuing, a company with ambitions that could finally meet Branson's stratospheric hype.

A space company, called Virgin Galactic.

WITHOUT A ROCKET or a spacecraft or any real knowledge of space travel, Branson registered the name "Virgin Galactic Airways," hopeful that he'd one day be able to start a space company.

He spent years talking to people in the space community, picking the brains of engineers "to see whether I could find anyone who would be competent to build spaceships," he later recalled.

But it just didn't seem possible. Space, it turned out, was exceedingly difficult—far more so than ballooning, or speedboats or airplanes. Years before, he'd had an opportunity to go to space. Mikhail Gorbachev, then the leader of the Soviet Union, had called to offer Branson what seemed just the sort of opportunity he had made his hallmark—this time, to become the first civilian in space.

However, accepting the invitation would cost Branson something like $50 million and require him to spend two years training in Russia. "I was building Virgin and just wasn't sure I could give up that amount of time," he said. "I was also just a bit worried that it would be perceived wrongly to spend that sort of money going to space."

So, for once, he did not say, "Screw it, let's do it." Instead, he said no—something that would sit as a "half regret" with him for years. "It would have been absolutely magnificent."

Meanwhile, the search for a rocket ship—or someone who could build one—was becoming frustratingly fruitless. By the early 2000s, "I had almost given up at that stage, moving on to other things."

Such as the Virgin GlobalFlyer. It was a sleek, single-seat airplane designed for yet another of Branson's adventures: to break the speed record for flying around the world. It was being built in the Mojave Desert by Rutan's Scaled Composites. When one of Branson's deputies went to check on it, he stumbled by accident onto something altogether different, something he knew his boss would want to know about immediately.

"You won't believe it, but I think we've got something even more exciting than the Virgin Atlantic GlobalFlyer here," he told Branson.

It was SpaceShipOne.

Knowing Branson would want to seize the opportunity, one of his executives rushed to register the name "Virgin Galactic"—only to find out that Branson had already done it years before.

Within days, Branson was in the Mojave Desert, looking at SpaceShipOne. This was it. This was the spacecraft that he'd been searching for for years. He had to be a part of it.

That day, he met with Paul Allen and Rutan at Rutan's house. The airplane inventor showed Branson his collection of paper napkins and scraps of paper on which he scribbled the ideas that popped into his head. Here were his concepts for spaceships and here was the idea for the "feather," the system that would detach the wings of the spaceplane from the body of the aircraft and fold upward, creating drag.

Rutan and Allen were thinking far ahead, indulging Branson's passion for space.

"We sat down and talked about hotels on the moon and day trips to the moon and all sorts of exciting things," Branson said. "And at the end of it we agreed that Virgin would help sponsor SpaceShipOne. And Paul and I agreed that if it was successful, we'd meet up and talk about trying to carry the program forward afterward."

SPACESHIPONE AND THE Ansari X Prize had shown that a commercial venture could put a person in space. The elusive dream had been realized, with Paul Allen as its benefactor. But Allen had never been comfortable with the enormous risks, and the fact that he was funding a venture in which people had such a good chance of dying was overwhelming. He had gotten through three flights that had made history. But, perhaps more important, no one had been hurt.

Branson visited Allen at his home in the Holland Park section of London. "I said, 'Look, I think there are hundreds of thousands of people who would love to go to space, and it would be a terrible waste if this was the end of it. We're going to develop a spaceship company, and obviously it would help if we could have the basic technology that's been developed to date.' And we shook hands on that. And bizarrely, I was the only person who went to see him about it."

After signing the licensing deal, Allen gladly turned over the keys to Branson, who was eager to take SpaceShipOne's simple design and blow it open to build an even bigger spaceplane. Whereas SpaceShipOne carried just a single person (though it carried the weight of three to meet the X Prize's requirements), Branson's SpaceShipTwo would carry two pilots and six passengers. A modest iterative step up this was not. Instead,

SpaceShipTwo would be a bold advancement, especially given how Space-ShipOne had bucked and rattled its way on its white-knuckled rides into space. But Branson didn't do incremental, or modest.

Despite the daunting task ahead, Branson wasted no time in injecting the full force of Virgin's vaunted PR machine into promoting the allure of the final frontier—and how he was going to make it available to the masses. Virgin Galactic would become the "world's first commercial spaceline," he crowed, one that would transform everyday paying tourists into full-fledged astronauts. He vowed that Virgin's maiden flight would happen as early as 2007, and that it would fly three thousand people in the first five years.

He even promoted Virgin Galactic during a Super Bowl commercial for Volvo in February 2005, just months after obtaining the rights to SpaceShipOne. The ad featured the liftoff of a rocket with a bumper sticker that said, "My other vehicle is a Volvo XC90 V8."

"Introducing the most powerful Volvo ever," the narrator said. "The Volvo XC90 V8."

As the narrator asked, "How powerful is it?" the camera cut to the astronaut inside the rocket, who lifted his visor to reveal that it was Branson.

"Powerful enough to get you into space," he said.

Branson treated space like a religion and he was its evangelist, preaching the virtues of space travel and the "life-changing" effects a trip into the stars could have, even if they lasted just a short while. Even if, at the time, they seemed little more than illusory. He was years away from developing a spacecraft capable of taking anyone to space, but that didn't stop the hype.

"We hope to create thousands of astronauts over the next few years and bring alive their dream of seeing the majestic beauty of our planet from above, the stars in all their glory and the amazing sensations of weightlessness and space flight," he said. "The development will also allow every country in the world to have their own astronauts rather than the privileged few."

Branson wasn't the first to try to sell the allure of space. During the 1960s, Pan Am started promoting trips to the moon as a way to cash in on the surging interest the Apollo program generated. So, it created a waiting list of passengers who wanted to go to the moon.

"We like to think of ourselves as pioneers," a Pan Am spokesman told the *New York Times* in 1969. "We were first across the Pacific and had many firsts across the Atlantic as well. We're going to be the first to fly the Boeing 747. So we would hope some day that we would be pioneering moon travel. That's why we keep the list."

Whether or not it was a PR stunt, Pan Am's customers bought it, signing up for their trips to the moon in droves. In return, the future astronauts received a letter addressed "Dear Moon First Flighter," signed by James Montgomery, Pan Am's vice president of sales.

"Thank you for your confidence that Pan Am will pioneer commercial Space travel, as it so often has here on Earth. We have every intention of living up to this confidence," the letter began.

It acknowledged that the "starting date of service is not yet known. . . . Fares are not fully resolved, and may be out of this world." The letter came with a card declaring the holder a "certified member" of Pan Am's "First Moon Flights Club," with a number denoted where the holder stood on the waiting list.

The *New York Times* reported that by early 1969, some two hundred had signed up. The list grew quickly, and agents at the airline grew used to handling the moon reservations with a matter-of-fact "For how many passengers, please?" On July 19, 1969, the day before the first moon landing, Pan Am's chief executive, Najeeb Halaby, told WCBS in New York that the airline was focused on "the concept of boosters that can be reused, of a space station which is like an airport in space, and frequent trips between the orbiting space station and various points on the moon."

By the time Neil Armstrong and Buzz Aldrin walked on the lunar surface, the reservation list had grown to twenty-five thousand names. By 1971, when it stopped taking the reservations, more than ninety thousand people had signed up, including Ronald Reagan and Walter Cronkite.

Pan Am folded in 1992, but believed up to the end that tourist trips to the moon were not just possible but an inevitability. "Commercial flights to the moon are going to happen," a company spokesman told the *Los Angeles Times* in 1985. "They might not happen next year, they might not happen in five years—but they will happen."

Now, years later, Branson was shooting for the edge of space, not the moon. But he approached the endeavor with the same gale-force enthusiasm, promising that flights to space were just around the corner.

"By the end of the decade, Virgin Galactic—the most exciting development in the story of modern space history—is planning to make it possible for almost anyone to visit the final frontier at an affordable price," the company said on an early version of its website.

Just after the final X Prize flight, and long before Virgin Galactic had anything close to resembling a new spacecraft, it invited potential customers to sign up on its website. By early 2006, Branson showed off a mockup spacecraft with a flat-screen television that gave a sense of what the journey into the heavens would be like. The seats were ergonomic, the windows numerous, and the first flights safe—and sublime.

The website laid out how the experience would unfold, in purple, at times creatively punctuated, prose. Seemingly describing an out-of-this-world, orgasmic acid trip, the narrative explained how the spacecraft would be tethered to the mothership, which would climb to 50,000 feet.

> Then the countdown to release, a brief moment of quiet before a wave of unimaginable but controlled power surges through the craft. You are instantly pinned back in your seat, overwhelmed but enthralled by the howl of the rocket motor and the eye-watering acceleration, which as you watch the read-out, has you traveling in a matter of seconds, at almost 2,500 mph, over three times the speed of sound.
>
> As you hurtle through the edges of the atmosphere, the large windows show the cobalt blue sky turning to mauve and indigo and finally to black. You're on a high, this is really happening, you're loving it and coping well. You start to relax; but in an instant your senses are back on full alert, the world contained in your spaceship has completely transformed.
>
> The rocket motor has been switched off and it is quiet. But it's not just quiet, it's QUIET. The silence of space is as awe inspiring as was the noise of the rocket just moments earlier. What's really getting your senses screaming now though is that the gravity which has dominated every movement you've made since the day you were born is not there any more. There is no up and no down and you're out of your seat experiencing the freedom that even your dreams underestimated. After a graceful mid-space summersault [*sic*] you find yourself at a large window and what

you see would make your hair stand on end if the zero gravity hadn't already achieved that effect. Below you (or is it above
you?) is a view that you've seen in countless images but the reality is so much more beautiful, so much more vivid and pro-
duces emotions that are strong but hard to define. The blue map,
curving into the black distance, is familiar but has none of the
usual marked boundaries. The incredibly narrow ribbon of atmosphere looks worryingly fragile. What you are looking at is
the source of everything it means to be human, and it is home.
You see that your fellow astronauts are equally spellbound, all
lost in their own thoughts and storing away the memories.

And best of all, the trips were right around the corner—just a few short
years away.

"In your dreams no more!" the company proclaimed.

It worked. People signed up, just as they had for the Pan Am trips to
the moon, plopping down $200,000 for their seat. Branson's version of
space was cool and sexy, with a Hollywood appeal, and by early 2006
Virgin Galactic already had $13 million in deposits. Brad Pitt and An-
gelina Jolie bought tickets. So did Ashton Kutcher and Tom Hanks and
Harrison Ford. But it wasn't just celebrities. Ken Baxter, a Las Vegas real
estate developer who signed up after watching a *60 Minutes* profile of
Burt Rutan, claimed he was the first customer.

But that distinction may very well go to Trevor Beattie, a London advertising executive, who was behind Wonderbra's "Hello boys" ad campaign that caused a stir in the United Kingdom. As a boy he'd been a fan
of space, and shortly before SpaceShipOne's final flight, Beattie called
Branson, whom he knew casually. Branson dared him to come to the
Mojave Desert for the last flight of the X Prize.

"And I said, 'Right. I'm going to call your bluff.' Instead Branson called
my bluff, saying, 'We're going tomorrow.' And I found myself on a plane
to L.A. I hadn't told anyone I was going. I just up and went."

After Brian Binnie's successful flight, Beattie vowed to be the first
to buy a ticket with Virgin Galactic. If he could do it, almost anyone
could. With a mop of near-shoulder-length curls, a pasty Englishman's
demeanor, and a slight paunch, he was about as far from an astronaut
as could be. But that's what he liked about it. Many of his fellow ticket

holders were "a group of self-selecting Herberts," he said, using cockney slang for "fool." "We're the opposite of the right stuff. We're the kind of wrong stuff. But so be it."

The first hundred to sign up were known as the Founders, and while other ticket holders would be allowed to put down a 10 percent deposit, they'd have to come up with the whole $200,000 up front.

IN EARLY 2007, Virgin broadcast in a press release, "Sir Richard Trains for Spaceflight," detailing how Branson was strapped into a centrifuge that spun around to simulate the g-forces of spaceflight. With him was James Lovelock, a scientist and author whom Branson admired for his work studying climate.

Branson had called him up and offered him a ticket. "My reaction was, 'Oh, I'm sure they'd never send me into space because in a few years I'll be ninety,'" he recalled. But Branson said that age was not a barrier: "My father is the same age as you, and he's going."

In the simulator, Lovelock recalled how a voice came on that said the spaceship was about to be released from the mothership. There was a countdown, "and you feel a jerk of it dropping off." Then, the voice came on again, this time to say that the rockets would fire soon. "And then you feel this enormous thrust and noise of the rockets firing, and then you feel the g-force come on. It's all rather fun."

Beattie trained by taking a ride in the "vomit comet," an airplane that flies in parabolas and gives passengers the feeling of weightlessness for a few minutes at a time. Floating through the cabin was an amazing experience—the feeling of flying.

It was extraordinary, right up until another passenger, a fat man, fell on him, breaking his toe. "I've had my first space injury," he said.

And a reminder of how when the gravity comes back on, things can come crashing down. The reality of getting to space was going to be a lot more difficult, and dangerous, than Branson made it appear. People were trusting him with their lives.

7

The Risk

THERE WERE SO many big egos—such outsize personalities
that could, like rocket fuel, combust—it was difficult to decide
who should sit next to whom at the conference room table. The
seating chart for this team of rivals—all big names in commercial space
and therefore competitors—was a delicate piece of social choreography.
The participants had to get along. They were trying to figure out how to
start a new industry.

At least they had a place to meet. Elon Musk had graciously offered to
host the meeting at SpaceX's El Segundo factory. Even though he was still
far from being a household name, he was known and respected by this
group. And his gesture gave the meeting credibility, making it easier for
people to RSVP in the affirmative. Getting everyone's calendar in sync,
however, was a nightmare. But, finally, they settled on a date—February
14, 2006.

Valentine's Day. Apparently romance isn't foremost on the mind of
these space tycoons, John Gedmark thought. Just out of graduate school,
the twenty-three-year-old intern at the X Prize Foundation had the un-
enviable task of arranging the seating in SpaceX's cramped conference
room.

Gedmark had sketched out the seating by hand on a yellow legal pad.
Elon was at the head of the table; as host, he earned that right. To his left
was Peter Diamandis, the organizer of the Ansari X Prize.

Robert Bigelow, the multimillionaire founder of Budget Suites of America, who wanted to build hotels in space, sat across from the Virgin Galactic representatives. John Carmack, the programmer behind such video games as Quake and Doom, sat near the middle.

Stu Witt, a former navy pilot who ran the Mojave Air and Space Port, was at the far end near Alex Tai and George Whittinghill, who were representing Richard Branson's Virgin Galactic, which was designing SpaceShipOne's successor, SpaceShipTwo.

Everyone who was anyone in the industry was here. Everyone, that is, except Jeff Bezos, or anyone from Blue Origin.

In 2006, Blue was still an obscure outfit, shrouded in secrecy, keeping many, even its industry brethren, at bay.

"It wasn't clear what their plans were," Gedmark said. "For all we knew, they were more of a small sort of R and D [research and development] shop."

Still, the members reached out to Blue Origin, inviting it to send someone to the meeting. But they couldn't get anyone to come out.

BURT RUTAN SAT at the far end of the table, still believing that the flights of SpaceShipOne were only the beginning, even though the famed spaceplane would never fly again.

After its historic flights, it was retired to the National Air and Space Museum, where it was put on display, hanging from the rafters between Lindbergh's Spirit of St. Louis and Yeager's X-1.

But while it would be preserved for future generations to see, the folks at the X Prize, and the industry in general, didn't want its enshrinement to signal the end of what they were trying to accomplish, which was nothing short of a viable industry that could take ordinary people to space.

Commercial space was having its "Lindbergh moment," the one giant leap Peter Diamandis and the people at X Prize hoped would spark a revolution in human space travel. Lindbergh had helped touch off a revolution in aviation, such that by 1955, more people were traveling on commercial airliners than taking the train. Lindbergh's flight had an immediate effect: ticket sales on commercial flights soared, as did the number of licensed aircraft.

IF THE INDUSTRY was going to have a real impact, the insurgency that had started with SpaceShipOne needed a second act. But some worried that the public's attention would soon wane, as it did after the Apollo moon landings. After achieving the impossible of putting a man on the lunar surface, NASA's human spaceflight program had struggled to recreate that magic.

Now after the Challenger and Columbia disasters, costing the lives of fourteen astronauts, many feared that the agency had become too bureaucratic and risk-averse to push humans deeper into space. That the drawdown after Apollo would become permanent. That Apollo would be an anomaly, a fluke, never to be repeated again. And that Eugene Cernan's hopeful promise—"We shall return"—made in 1972 as he became the last man to walk on the moon, was eroding from prophecy to fallacy.

This was the country of Neil Armstrong and Chuck Yeager, the Wright Brothers and Lewis and Clark. Opening frontiers has long been part of the American DNA, from the *Mayflower*, to Manifest Destiny, to the moon. Musk saw the discovery of adventure as an inherently American ideal.

"The United States is a distillation of the human spirit of exploration," he once said. "Almost everyone came here from somewhere else. You couldn't ask for a group of people that are more interested in exploring the frontier."

If NASA, or Congress, or any president wouldn't stand up as John F. Kennedy did in 1961 when he promised to send a man to the moon within a decade, then this class of entrepreneurs would attempt it.

Instead of hoping Kennedy would rise from the grave and give them the space program they wanted, maybe they were, themselves, the people they'd been waiting for.

THEY WERE AT SpaceX's headquarters to officially band together and call themselves the "Personal Spaceflight Federation." They wanted their movement to catch fire and spread, and thought that the industry needed to form an industry association, to keep the momentum going and show Washington, DC, and the Federal Aviation Administration (FAA) that they were for real.

Like Musk, several of the members were Silicon Valley types, and so "personal" was chosen to mimic the "personal computer." They wanted to signal that just as computing went from industrial mainframes to small desktops, so would space, too, become an individual experience.

In addition to that quixotic goal, there was a real immediate concern. SpaceShipOne's thrilling, white-knuckled flight not only captured the attention of the world, but Congress and the FAA as well. As some in the industry had feared, the federal government was now weighing how to best regulate this emerging industry.

To the group, many of whom had a libertarian bent, the words *congressional oversight* and *federal regulation* were, at best, anathema to their core beliefs. At worst, such government involvement could lead to the demise of their companies. By speaking with one voice, they could help craft the regulations, ensuring that Washington didn't stifle a fledgling industry before it had left the nest.

In preparation for the Valentine's Day meeting, Gedmark realized that the Personal Spaceflight Federation was really a federation in name only. Yes, the group had put out a press release announcing its formation. But it didn't have any money or legal standing as a nonprofit organization. Gedmark knew it would need both.

He took care of setting it up as a nonprofit, getting the California secretary of state's office to certify its articles of incorporation a week before the Valentine's Day meeting. And he put together a draft memo for his boss that outlined the regulatory hurdles the industry could face.

It began: "The Personal Spaceflight Federation is a non-profit organization, incorporated in the State of California, dedicated to resolving the regulatory, legal, political and broad strategic challenges the personal spaceflight industry faces moving forward."

Gedmark's memo warned that "the danger of overly burdensome regulations continues to be a significant risk. Almost as critical is the danger of muddy, chaotic, or inconsistent regulations, as an atmosphere of uncertainty or chaos in an industry can quickly dry up badly needed sources of capital."

He acknowledged the difficulties of the marketplace they were trying to disrupt, writing, "The entrenched aerospace industry is not only increasingly monopolistic, but is heavily subsidized by the federal government."

He outlined plans to develop "informed consent" standards, an attempt to have the industry treated like a thrill sport, the same as bungee jumping or skydiving—if you're crazy enough to jump out of a plane, be

Elon Musk unveils the version of the Dragon spacecraft designed to fly astronauts at an event at SpaceX's headquarters, 2014. *Courtesy of NASA/Dimitri Gerondidakis.*

Jeff Bezos shows off Blue Origin's crew capsule and the New Shepard booster at a conference in Colorado Springs, 2017. *Courtesy of Christian Davenport.*

A Falcon 9 lands on a ship in the Atlantic Ocean, 2016, after hoisting a commercial satellite to space. *Courtesy of SpaceX.*

A Falcon 9 booster stage arrives at Port Canaveral, Florida, after landing on a ship in the Atlantic Ocean, 2016. *Courtesy of SpaceX.*

Blue Origin's New Shepard booster launches from the company's site in West Texas, April 2, 2016. *Courtesy of Blue Origin.*

Brian Binnie *(left)*, Paul Allen, and Burt Rutan stand in front of
SpaceShipOne after Binnie successfully flew the spacecraft in 2004,
winning the Ansari X Prize. *Copyright © Mojave Aerospace Ventures
LLC; courtesy of Scaled Composites.*

Paul Allen speaks as SpaceShipOne
goes on display at the National Air
and Space Museum, Washington, DC,
2005. *Copyright © 2005, Larry Morris/
Washington Post.*

Elon Musk gives President Barack
Obama a tour of SpaceX launchpad
at Cape Canaveral Air Force Station,
2010. *Courtesy of NASA/Bill Ingalls.*

Lori Garver, then the NASA deputy administrator, tours Blue Origin's facility in 2011, meeting with members of the company, including Jeff Bezos. *Courtesy of NASA/Bill Ingalls.*

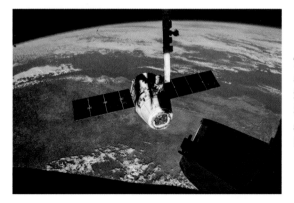

The SpaceX Dragon spacecraft being released from the International Space Station, 2014, after a cargo delivery mission. *Courtesy of NASA.*

Gwynne Shotwell, SpaceX president and chief operating officer, speaks in front of launch complex 39A at the Kennedy Space Center, 2017. Also shown are Kennedy Space Center director Bob Cabana *(left),* and Tim Hughes, SpaceX's senior vice president and general counsel *(right). Courtesy of NASA/Kim Shiflett.*

Jeff Bezos and Buzz Aldrin at the National Air and Space Museum, 2016, after Bezos was awarded the Heinlein Prize, an honor named for the science fiction writer that comes with a sword. *Courtesy of the Heinlein Prize Trust.*

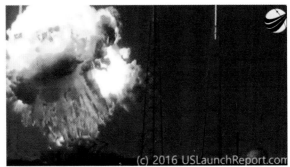

A SpaceX Falcon 9 rocket explodes during fueling ahead of an engine test at Cape Canaveral Air Force Station, 2016. Mike Wager/*US Launch Report.*

NTSB Chairman Christopher Hart and NTSB investigators with Virgin Galactic pilot Todd Ericson after the 2014 fatal crash of SpaceShipTwo in the Mojave desert. *Courtesy of the NTSB.*

Richard Branson celebrates with a giant leap while showing off Virgin Galactic's new SpaceShipTwo in Mojave in 2016. *Courtesy of Virgin Galactic.*

Richard Branson unveils the new SpaceShipTwo, dubbed Unity, at an event in Mojave, California. *Copyright © 2016, Ricky Carioti/Washington Post.*

Virgin Galactic's SpaceShipTwo in a glide test flight over Mojave, 2016. *Courtesy of Virgin Galactic.*

Blue Origin paints a turtle, the company's mascot, on its booster after every launch. *Courtesy of Blue Origin.*

A model shows off SpaceX's spacesuit. *Courtesy of SpaceX.*

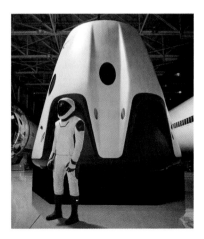

Paul Allen's Stratolaunch, coming out of its hangar in Mojave, California, 2017, would be the largest airplane ever flown. It is designed to "air launch" as many as three rockets. *Copyright © Stratolaunch Corporation.*

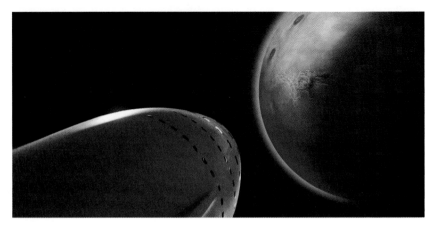

An artist's rendering of what SpaceX's BFR rocket would look like approaching Mars. *Courtesy of SpaceX.*

our guest, but know that death is one of the possible outcomes. Just remember to pull the rip cord.

Finally, the industry also had to prepare for the worst, Gedmark warned.

"Unfortunately, the personal spaceflight industry must proceed assuming that a fatal accident is inevitable," he wrote.

DEATH WAS MORE likely a "when, not if" outcome, an unavoidable fact that they should confront and plan for. But death shouldn't stop them. It shouldn't get in the way. Progress was not possible without it. That was true in space as it was in all manner of expeditions, from crossing the Atlantic, to exploring the poles, to opening up the West.

When Ernest Shackleton set off to cross Antarctica in 1914, he was said to have placed a newspaper advertisement that read: "Men wanted for hazardous journey, small wages, bitter cold, long months of complete darkness, constant danger, safe return doubtful. Honor and recognition in case of success." (Some have cast doubt as to whether this advertisement was ever actually placed. Still, it was clearly a harrowing and dangerous journey.)

Similarly, an avalanche center warning to the backcountry skiers who flock to Tuckerman Ravine, the glacial cirque just below the peak of Mount Washington in "Live Free or Die" New Hampshire, read: "Visitors to the Ravine should never come expecting to be rescued when something bad happens. Don't rely on other people being around to help you; ultimately your party may be the only rescue team available to respond."

As a guidebook pointed out, Tuckerman wasn't just a ravine but a culture clash—it subscribed to an ethos that was "anathema to many of the values of modern society. It takes hard work to get there, there are no rules, dire consequences can follow from mistakes and you have nothing to show for your courageous efforts save for a fleeting track in the snow."

The act of embracing that death was a likely outcome in the quest to open up the space frontier might seem like a macabre exercise. But it was liberating and, in a way, even optimistic, a view beyond the grave to a point on the horizon where the sacrifice would be worth it. Forays into the unknown had to be met with a steely mix of thorough preparation and blind hope, like Magellan's crossing the strait through the southern tip of Chile that would one day be named for him, and entering the

Pacific for the first time, unaware of how vast it stretched out ahead of him, not knowing when he'd hit land.

Mike Melvill had escaped death during his harrowing SpaceShipOne flights—one flying blind, the other in an uncontrolled spin. But he had held on and gutted it out, and in the process had earned the glory, the "honor and recognition" that Shackleton was said to have promised nearly a century earlier.

In modern society, there were few places that allowed such freedoms—duly warned, but uninhibited. Nobody told Lincoln Beachey, one of the barnstormers in the early days of aviation, that he couldn't break the altitude record of 10,466 feet by strapping on an extra 10-gallon tank to his Curtiss D plane, flying until it ran out of fuel and then gliding back to the ground. No one told him he couldn't fly over the edge of Niagara Falls as if he were part of the waterfall itself, getting so close to the bottom that it appeared his plane had been lost in the misty whirlpools below until he pulled up in a dramatic escape while being watched by a crowd of 150,000. And no one told him he couldn't perform the dangerous bit of aerial acrobatics that ultimately killed him—flying in a vertical S that ended up with a crash in the San Francisco Bay in 1915.

Lots of pilots were pushing the edge at the time, seen not just as daredevils, but also—and especially by those hoping to open up space—as martyrs to a greater cause, one worthy of sacrifice. By definition, exploration—a foray into the unfamiliar—demanded a high tolerance of risk, a willingness, as Joseph Conrad wrote in Lord Jim, to "in the destructive element immerse."

Outside of the battlefield, there was perhaps no more destructive element than being strapped in on the top of a rocket, a controlled explosion of highly flammable propellant. There is a reason why NASA chose so many of its astronauts from the US Navy and Marine Corps, courageous battle-tested fighter pilots all with the Right Stuff.

Astronauts and test pilots talk openly of death, for the same reason marines talk openly about killing—to desensitize it, make it real, and eventually come to accept it as the inevitable fact of life. Their will had been written and signed long ago. When they'd lose their life—at a ripe old age, or triggered far too early by a sniper's bullet or a blown rocket engine—was a risky game of Russian roulette, a matter of chance.

Before he was killed in the fire that engulfed the crew of Apollo 1 on the ground during a preflight test, Gus Grissom had prepared himself for an outcome he knew was highly probable.

"If we die, we want people to accept it," he said. "We're in a risky business, and we hope that if anything happens to us it will not delay the program. The conquest of space is worth the risk of life."

But that was then. The swashbuckling era of Mercury and Apollo had passed, like poor, lonesome cowboys riding off into the sunset. It was replaced by the solemn rectitude of the parents who spent too much time at the funerals of their offspring, who knew the consequences and were rightfully chastened and scared.

During the Apollo 11 mission to the moon, the average age in mission control was just twenty-six. Gene Kranz, the flight director with the flattop and steely nerves, was thirty-five, the senior statesman, "the old man in the room," he said.

They were young and invincible, full of so many romantic illusions that they didn't know that the task President Kennedy had given them was impossible.

Since then, NASA had continued to pioneer, sending rovers to Mars and robots that scoured the far reaches of the solar system in one amazing feat after another. The Hubble Space Telescope had unlocked mysteries of the universe. Forty years after it was launched, the Voyager 1 spacecraft had reached interstellar space, traveling more than 13 billion miles from the sun. Voyager 2, also launched in 1977, is the only spacecraft to fly by all four of the outer planets of Jupiter, Saturn, Uranus, and Neptune. Both continued to communicate with NASA daily. The Cassini spacecraft became the first ever to orbit Saturn, while making new discoveries about its mysterious rings and moons.

But nothing quite had the cachet or the thrill that came when a human being was aboard the rocket.

Decades later, as NASA's average age grew to nearly fifty by the height of the shuttle era, its aversion to risk grew. After the Challenger disaster killed all seven on board; then Columbia, another seven, the investigations and accusations piled up and the youthful invincibility was gone.

Now the commercial space industry that was trying to pick up where NASA had left off feared that if "crew were lost" in an "anomaly"—the formal, anesthetized language used for "people" and "killed" and "explosion"—then there would be real trouble. Congressional and FAA investigations. Subpoenas and reports and hearings. All of which could bring the federation's nascent efforts crashing down.

THE FREEDOM TO kill yourself in all manner of stupid ways was part of the American way, and it's partly what made it appealing to Musk, a South African immigrant drawn to the United States for its free markets, its can-do spirit, and its entrepreneurial bent. He had moved to California, by way of Pretoria, Ontario, and Philadelphia, finally out west, following the Silicon Valley gold rush.

Musk had always had a bit of wanderlust, asking his father as a young boy, "Where is the whole world?" He came from a family of adventurers. His maternal grandparents, Joshua and Wyn Haldeman, had emigrated from Canada to Pretoria to escape what they considered a repressive political climate but also seeking "a base for exploration," Musk said.

Haldeman wasn't a barnstormer but an accomplished amateur pilot who was handy; he flew all over North America and Africa and Asia, and once in 1952, on a 22,000-mile journey across the globe. He also was believed to be the first person to fly from South Africa to Australia in a single-engine plane, and as Musk pointed out, "he did this in a plane with no electronic instruments. In some places they had diesel and in some places they had gasoline and so he had to rebuild the engine according to whatever fuel they had."

An inspiration to his grandson, Haldeman, who was born in Minnesota, was also an "amateur archaeologist," Musk said, who was fascinated with the Lost City in the Kalahari Desert. He made a dozen expeditions there with his children, including Musk's mother, Maye, in tow.

They were searching for the mythical city supposedly discovered by Guillermo Farini in the late 1800s. Starting in 1957, Haldeman retraced the explorer's steps, into territory rarely visited, using maps with little information, while sometimes writing their own. Year after year, Haldeman plunged into deep, barren country, sometimes flying just a couple hundred feet aboveground so that his guide could navigate by studying the landmarks on the ground.

"There is something particularly fascinating about traveling through country which is unknown, untamed and untouched by man," he wrote.

The family brought tents but rarely used them. Their guide slept by the fire feet first, so that when "his feet got cold he knew it was time to put some more wood on the fire," Haldeman wrote. However, Lee, the youngest child, who was four on his first trip, did sleep with a roof over his head—"on the front seat of the car, as he is too tempting a morsel for any hungry night prowlers."

Those included all sorts of predators, including leopards and a pair of lions, which Haldeman once bumped into, almost inadvertently, and backed away from slowly, saying to his wife, "Look, Wyn, a lion." They scared them away with frantic shouts and a torch until the lions "went up on to the hill and watched us until dawn."

Musk was three years old when his grandfather died, so "my only exposure to it was my grandmother showing slideshows of the various adventures," he recalled. "When I was a kid I found the slideshows kind of tedious, but maybe it stuck in some way. Now I'd like to see the slideshow. But as a kid I was, like, 'I want to go play with my friends. Why are you showing me these slides of the desert?'"

In founding SpaceX, Musk believed that in addition to trying to make humans a multiplanetary species—with the ultimate goal of sending people to Mars—he saw space travel as the greatest adventure ever, even greater than the quixotic searches for the Lost City.

Although there was, as he said, the "defensive reason" to go to Mars to colonize another planet—so that humanity would have another place to go in case anything happened to Earth—this was not what inspired him to go to Mars.

"The thing that actually gets me the most excited about it is that I just think it's the grandest adventure I could possibly imagine. It's the most exciting thing—I couldn't think of anything more exciting, more fun, more inspiring for the future than to have a base on Mars," he once said. "It would be incredibly difficult and probably lots of people will die and terrible and great things will happen along the way, just as happened in the formation of the United States."

Just as his grandfather had been free to take off in his airplane and go wherever he wanted—the Kalahari Desert, Australia, South Africa— Musk also enjoyed the thrill, and risks, of flying. For a short while he even owned a Soviet L-39 fighter jet. "I'd do acrobatics in it and fly at tree top level, fly up a mountain invert and fly back down the other side,"

he said. "But then I was like, man, this was made by some Soviet technician and *maybe* they tightened the bolt right, or maybe they didn't. Not
a lot of redundancy. It was like, 'This is crazy. I've got kids. I have to stop
doing this.'"

He felt that humanity's exploration of space should be as unencum-
bered as the opening of thousands of other frontiers, from the bottom of
the ocean to the peaks of mountains.

Once, in the early days of SpaceX, Musk asked a space industry executive: "Do you know how many people have died on Mount Everest?"

A few hundred. Many of the bodies lying there, entombed, frozen reminders of the perils of exploration all the way to the top.

THE GOVERNMENT REGULATORS were already circulating, and some
members of Congress were looking askance at this new industry that
wanted to be able to fly people into space with little oversight.

A year before the Valentine's Day meeting, a congressional hearing
titled "Commercial Space Transportation: Beyond the X Prize" gave the
members of the federation a jolt. James Oberstar, a longtime member of
Congress, said he was "watching this process like a hawk."

While he said he was "a convert to the cause" after once viewing
commercial space "as, quite frankly, a distraction," he also called for
more robust regulations, ones that would not only protect the people
on the ground, but the passengers who had chosen to fly. The FAA had
what he called a "tombstone mentality—wait until someone dies, then
regulate. . . .

"That is not safety," he continued. "That is being reactive, and that is
what offends me."

Under its regulations, the FAA protected the "uninvolved public and
property" on the ground. But it offered no such protection to the actual
passengers of the aircraft. Oberstar thought this was ludicrous, and that
it needed to be changed.

"We ought to worry about the people on the plane," he said.

But others at the hearing praised the accomplishments of the X Prize
and the enthusiasm it generated. Congressman John Mica said that the
flight of SpaceShipOne had "launched a whole new era in space." The flight
heralded an exciting future and "altered our vision of what the aviation
system of the future will look like," he said. "We now see the possibilities,

including the development of space tourism, US spaceports, rapid global transportation."

The testimony of the FAA administrator, Marion Blakey, was perhaps most important. If she called for Congress to crack down on the enterprise, the companies in the Personal Spaceflight Federation could be in real trouble.

Instead, she came out strongly for the industry, equating the moment of commercial space to where commercial aviation was a hundred years earlier, when the Wright Brothers took their first flight at Kitty Hawk. She applauded the efforts of entrepreneurs, such as Musk, Branson, and Allen, who were betting their fortunes on the industry, calling them "astropreneurs."

"The space you and I grew up knowing dealt largely with final countdowns, and Jules Bergman," she said, referring to the ABC news broadcaster who covered the space program during the 1960s. "Space was a place where you saw flickering black-and-white photographs, images with leaps of mankind. Not anymore. America's love affair with space is vicarious no more. There is a bold new group of people—astropreneurs—and their aim is to bring space flight into everyone's grasp."

Coming from the head of the nation's aerospace regulatory agency, it was a strong endorsement that gave the entrepreneurs reason to feel optimistic. They could, at least for a moment, breathe a sigh of relief.

But for all the enthusiasm for this new industry and the future it heralded, there were, however, tough questions about how it should be regulated.

In her testimony, Blakey confessed that keeping up with the fast-moving industry "is going to be a real challenge."

The Personal Spaceflight Federation had a pair of representatives on the hearing panel, ready to push back against calls for what it perceived as cumbersome rules. Michael Kelly, a member of the federation who also served on an FAA advisory committee, said that they were in unprecedented territory. No one had ever tried to fly into space commercially. And therefore no one had tried to regulate it. If the government came in too forcefully it "would be tantamount to prohibiting personal spaceflight as an activity," he said.

Instead, the industry needed to come up with its own standards, ones that it would develop as it gained experience, bit by bit. "The only people who are gaining the experience that can be applied quickly and in the time required to support this industry are people who are in it them-selves," said Kelly.

Will Whitehorn, the president of Virgin Galactic, was also at the con-gressional hearing, and he pointed out that killing your customers is gen-erally not a good business practice.

"Given that we have had eighteen hundred people who have now ap-proached us wishing to fly in the early years, and given that they read like a textbook of Hollywood, Congress itself, international stardom, we are hardly likely to launch space flights which will kill these people," he told the committee. "It will not be our intention to operate in anything but the safest way possible."

The hearing went as well as they could have hoped. But they'd have to be vigilant.

AFTER THE MEETING on Valentine's Day adjourned, Musk offered to give the group a tour of his facility. To this group of engineers and entrepre-neurs, it was like an invitation to a six-year-old to visit a chocolate factory.

As Musk guided them through the factory floor, the group "let loose with detailed, technical questions, and he answered all of them," Ged-mark said. "Not once did he say, 'I don't feel comfortable answering that because it's proprietary.' . . . It was certainly impressive."

At one point, John Carmack, the video game programmer who had started a rocket company, wandered off on his own, curious about a wir-ing diagram splayed out on a table. After studying it intensely, he looked up at Musk and said, "I have a question. What gauge of wire did you use right here?"

With that, Musk, who had been taking detailed, rapid-fire engineering questions, was finally stumped.

The team was already working on the company's next rocket—the Falcon 9. (It had abandoned plans to build the five-engine Falcon 5 that Musk had promised years before at the FAA reception.)

But something was missing from the factory. The very first rocket that SpaceX would attempt to fly was down at a launch site thousands of miles away, in the Marshall Islands of the Pacific Ocean. The maiden launch had been delayed and delayed. But the company was now getting close to its first attempt at lighting its engines.

SpaceX couldn't be sure, though, whether it would fly or explode.

8

A Four-Leaf Clover

O N JULY 29, 2003, the Defense Advanced Research Projects Agency (DARPA) issued an announcement that it was looking for a "transformational capability." Which was not unusual since the mysterious Pentagon agency was always looking for transformational capabilities. But even by DARPA standards, this solicitation—for a space weapon—stood out.

The Pentagon wanted to develop "a means of delivering a substantial payload from within the continental United States (CONUS) to anywhere on Earth in less than two hours," the announcement read.

This was no benign delivery service, however. The "substantial payload" would be a deadly arsenal—missiles, bombs, and a mysterious spacecraft capable of traveling at hypersonic speeds, or at least Mach 5, five times the speed of sound, some 3,800 mph. Launched from the East Coast, the strike would be able to hit Baghdad in just over an hour and a half.

Like many Pentagon programs, this one was given a clunky acronym, FALCON—which stood for "Force Application and Launch from CONUS." And it was born from necessity, or at least a wish list coming out of the top reaches of the Pentagon.

Although United States forces were able to hit Afghanistan with a punishing wave of ordnance in the wake of the September 11, 2001, terrorist attacks, and then light up Baghdad in 2003 with a devastating "shock and awe campaign," those attacks required a substantial buildup of forces in the region, which consumed precious time in the heat of war.

"While advancements in target identification and precision strike have been abundantly demonstrated, deficiencies in engaging and defeating time-critical and high value, hard and deeply buried targets (HDBT) have also been revealed," the FALCON solicitation read.

In other words, the Pentagon needed to act fast if it got word that someone such as Osama bin Laden was hunkered down in a bunker somewhere.

Now the Pentagon was looking for a way to strike targets thousands of miles away without having to rely on forward operating bases or aircraft carriers. The way to do that was to go into space.

"It was clear to leadership at the Pentagon that we had no quick way to reach out and touch somebody, so to speak—a Saddam Hussein or somebody that we needed to take care of quickly," said Steven Walker, then FALCON's program manager. "If you didn't have bases already prepositioned close to where action needed to occur, there's no quick way for the US to respond."

Developing a system that could be up and running quickly and hit anywhere in the world within a couple of hours was just the sort of near-impossible challenge that DARPA took on all the time. It was, as they liked to say at the agency, "DARPA-hard." Since its creation in 1958, when Bezos's grandfather, Lawrence Gise, was hired, DARPA had come a long way. Although Gise was worried that the agency would go "up in blue smoke" because of the threatening political pressure, it instead solidified itself as an indispensable, if mysterious, part of the defense establishment. With a relatively small budget, it advanced all sorts of military technology, with the goal of staying a step or two ahead of the enemy and preventing another Sputnik-like surprise.

DARPA was tasked with looking into the future to envision what sorts of technologies the United States would need for the future of war: "To cast a javelin into the infinite spaces of the future" was its motto, a quote from Hungarian composer Franz Liszt. Walled off from the rest of the giant Pentagon bureaucracy so that it could innovate freely, the agency strove for nothing short of revolutionary advancement and "engineering alchemy" that would pierce the realm of science fiction. It had been given the authority to hire as it needed, as it sought "extraordinary individuals who are at the top of their fields and are hungry for the opportunity to push the limits of their disciplines."

During Gise's time, DARPA, then known as ARPA, was focused on preventing nuclear war and winning the space race. It even helped develop NASA's Saturn V rocket, which took the Apollo astronauts to the moon. Since then, its reach and influence had broadened. In the late 1960s it started work on what would become ARPANET (Advanced Research Projects Agency Network), a network of computers in different geographic locations that became a precursor to the Internet.

Over the years, it helped develop all sorts of technological advancements that have transformed war, and, in some cases, everyday life. DARPA helped give birth to the Global Positioning System (GPS), stealth technology, cloud computing, early versions of artificial intelligence, and autonomous aerial vehicles. As early as the late 1970s, it was working on a "surrogate travel system" that created something like a Google Street View map of Aspen, Colorado. More recently, its work was focused on underwater drones, geckolike gloves designed to enable soldiers to climb walls, humanoid robots, bullets that can change direction, and a blood-cleansing "artificial spleen" to help treat sepsis.

FALCON, if it could be achieved, would join the pantheon of "breakthrough technologies" fostered by the agency. The program came in two parts. The first was to develop the hypersonic vehicle that eventually would be able to take off and land from a military runway and then hit anywhere in the world within a couple of hours.

In the near term, though, the hypersonic spacecraft would be launched from a rocket. And that presented a problem, as the FALCON solicitation pointed out: "Existing booster systems are costly and in limited supply."

So, the second part of the program sought a new kind of rocket, one that was inexpensive and could launch quickly—"after authorization from an alert status within 24 hours." It would have to be able to carry not only the hypersonic weapon, but also small satellites that could be used for spying. And it would have to be cheap—less than $5 million a launch.

ONE OF THE proposals that DARPA, and its partners in the US Air Force, received was intriguing, not just because the name of the rocket, Falcon, mimicked the name of the program, but because it was already well under development. Walker, the program manager, had never heard of the company SpaceX, or its founder, Elon Musk. But once he saw Musk's plan to develop a rocket that could launch extremely inexpensively, he wanted to know more about this Internet tycoon turned space entrepreneur.

"We were interested in bringing him into the program because his development was obviously in the early stages, but it looked like he was going down a path that would enable affordable launch," recalled Walker, who would go on to be the acting director of the agency. "Our goal was five million dollars. He was saying six million, which was much lower than anyone was doing it."

When Walker visited SpaceX's headquarters, he liked what he saw: a dedicated and passionate band of rocketeers that "had basically developed the first American-made engine in a long time," he said. "I was impressed with what they had done technically, and with Elon's business model and plan. I was really impressed with his ability to bring in key people on his team and give them the resources early on to go make this thing happen."

He was also struck by how they were building the rocket almost entirely on their own, without subcontractors. "Elon cared so much about the quality of rocket," that he'd pay for it to be made in-house, Walker said.

Walker was convinced. In 2004, DARPA agreed to invest a few million dollars into SpaceX, helping fund its first launch attempt. After years of relying on Musk's fortune, the company finally had some outside backing, if a minimal amount.

Now all it had to do was fly.

THE LAUNCH WAS supposed to happen at Vandenberg Air Force Base. SpaceX had spent $7 million on a pad there, adapting it for its Falcon 1 rocket. But in 2005, some of the traditional contractors, including Lockheed Martin, complained about SpaceX's presence at Vandenberg. Lockheed was worried that if SpaceX's rocket exploded, it could damage the facilities at a time when Lockheed was working to launch sensitive military satellites worth millions.

The air force had promised to help SpaceX find another suitable accommodation, but it became clear to Musk that the offer was mere lip service. The US military had a long cozy relationship with Lockheed and was comfortable with the defense contractor's Atlas V rocket. SpaceX was new to the neighborhood, an outsider with an unproven rocket built in what was essentially a garage in El Segundo.

Instead of staying on the range, SpaceX moved to an outpost in the Marshall Islands in the middle of the Pacific Ocean, thousands of miles away. DARPA helped with the transition, but Musk was fuming once again at the large contractors—and the air force, which he felt was screwing him.

"It's like you build your house. . . . somebody else builds a house next to you and tells you to get out of your house," he said at the time. "Like, what the hell . . . after we've just made that big investment and everything. We're going to fight that issue because it is just fundamentally unfair."

Just because the Atlas V was needed for national security launches, "that doesn't mean they can completely shaft us," he said.

The distance wasn't the only problem with the site in the Marshall Islands. Musk worried about the conditions wreaking havoc on his new rocket. "I don't think there's a place in the world with more corrosion," he said. "It's the perfect environment of right temperature, humidity, and salt spray."

There were, however, some upsides. While the range, known as the Ronald Reagan Ballistic Missile Test Site, was a government facility, SpaceX and DARPA had almost free rein with limited interference. Surrounded by turquoise water, sandy beaches, and palm trees on an island known as Kwajalein Atoll, or Kwaj, the setting felt like an island vacation spot.

"It was fun; we had the run of the place," Walker said. It was a location where the young company "could learn and practice" the precise art of the launch.

The teams slept in army-style barracks, often two to a room. There was outstanding snorkeling. Fishing boats were available to rent for ten dollars an hour. And there was lots of grilling out on the beach at sunset. "They say the only things to do on Kwaj are: work, sleep, work out, fish, drink, and screw," said an air force official who was there. "One friend who worked there for two years told me that he knew every square inch of every available woman on the island under sixty."

Musk's team worked to get the site ready as if they were on a mission. With little else to do on Kwaj, they kept extraordinarily long hours that both impressed and worried the small band of government officials along with them, who wondered whether the staff would burn out before the rocket fired.

"That was a big question in my mind—how could they keep that pace?" said Dave Weeks, a NASA official on loan to DARPA to help oversee the launch.

There was a cordial but uneasy relationship with Musk and the government team, especially at first. Controlling and detail-oriented, Musk wasn't particularly interested in their insights or suggestions. It was his rocket—not theirs—and he was suspicious of outsiders.

Weeks could understand why. Weeks had met Musk in 2002 or early 2003, when he and Gwynne Shotwell, who would later become the president of SpaceX, visited NASA's Marshall Space Flight Center. They weren't long into their meeting when a fire alarm sounded and they had to evacuate the building. Outside, a security guard approached as they waited to go back in, and said that he'd been informed that Musk, a green card holder born in South Africa, hadn't gone through the required background check the center required of foreign nationals.

The spaceflight center sat on the US Army's Redstone Arsenal, and, escorted by security, they went to the officers' club there.

"I thought this was not a great start," Weeks recalled. Musk was polite and gracious, but "miffed a bit," said Weeks.

Essentially, Musk had been kicked off the Marshall Space Flight Center. Then, he was kicked off Vandenberg Air Force Base. And here was the government, working with them side by side on Kwaj. Musk remained polite and calm. But he kept a distance. To the greatest extent possible, SpaceX was going to do this on its own.

He "wasn't high on government help," Weeks said.

ON MARCH 24, 2006, just four years after Musk had founded SpaceX, the seven-story-tall Falcon 1 rocket stood on the pad on Omelek Island, a couple miles north of Kwaj, ready at long last to launch. The flight had been delayed repeatedly over a period of months, and frustration was mounting, as were doubts about whether the rocket would ever get off the ground. But Musk was determined. He had fought so hard to get here, invested so much of his money. He had battled with the government, the entrenched contractors, and had pulled off stunts to get noticed by NASA, hoping they'd see that his company was for real.

Nothing would prove that, and quiet his critics, like a successful launch. Now there was something else on the line. In early 2006, NASA announced that it was starting a program to help commercial companies like SpaceX build rockets and spacecraft that would eventually be able to fly cargo and supplies to the International Space Station.

With the space shuttle set to retire in the coming years, NASA, under the administration of George W. Bush, was making a bold bet—that the private sector would be able to provide a taxi-like delivery service to the station, an orbiting laboratory, some 250 miles high. If the commercial sector could take over the relatively routine operations in what is known as low Earth orbit, that would allow NASA to do the hard stuff, to explore in deep space.

Michael Griffin, the NASA administrator at the time, believed that by investing in a few companies to help them develop their rockets and spacecraft, NASA could touch off an industry of viable commercial rocket companies. NASA, then, would become a customer, buying a service to launch supplies to the space station.

"Commercial enterprises historically, if they can succeed at all, they're cheaper than government enterprises, more dynamic," he said.

Griffin wanted just to provide "seed money," and was concerned that if "you have a lot of government money into the enterprise, then it becomes a government enterprise. And that's what I wanted to avoid."

He carved out a $500 million proposal in his budget, called the Commercial Orbital Transportation Services program, or COTS, that would be divided by two or three companies. How did he come up with that amount? "Truthfully, I just made it up," he said. It was enough to be significant for a startup. But in the space business, a couple hundred million dollars can burn up quickly, so the companies would have to come up with funding of their own.

For SpaceX, winning a spot in the COTS program wouldn't just be a much needed stamp of endorsement from NASA, but the investment would provide a measure of stability. SpaceX had to be considered a long shot. First, it had to convince NASA officials that the business was going to be around in a few years. NASA didn't want to invest in a company that would evaporate overnight.

NASA wanted to know "who would our potential investors be, how much money did Elon have to invest?" Shotwell recalled.

Then there was another question about the capabilities of a company that had never launched anything, yet was talking about one day going to Mars. "Can this very young, brash company do what they say they were going to do?" she said.

While it was preparing to launch the Falcon 1 from Kwaj, SpaceX crews in California were working on the next rocket, the much more powerful Falcon 9, which could be used in the COTS program.

Virtually everyone who wasn't at Kwaj was assigned to the Falcon 9 program. It was all hands on deck. Starting as soon as it received the request for proposals, the company was "largely shutting down and everybody working on this particular effort," Shotwell recalled. "At the time, it was that critical to the company and to our future."

The competition put even more pressure on the crews in Kwaj, who had delayed the launch again and again through 2005 for various technical issues. But now they were finally ready to go.

To launch a rocket successfully on the very first try would be an enormous coup. The history of rocket development is filled with blooper reels of rockets blowing up every way possible, on the pad, just above the pad, or going way off course, spinning wildly like a balloon losing air until they come crashing down in a fireball.

"A million things have to go right in order to have a successful launch, literally, and only one thing has to go wrong to have a really particularly bad day," Shotwell would later say.

Leading up to the launch, Musk acknowledged the possibility of failure, telling a reporter that the company could withstand a failed launch or two. "If we have three consecutive failures . . . it's not clear to me that we know what we're doing and maybe we should go out of business."

Sitting in the control room, Walker, the DARPA program manager, had another goal for the launch: "I was hoping no one was going to get hurt."

THE MOMENT HAD been four years in the making, and it all came crashing down in less than one minute. The count from T-minus 10 seconds seemed to proceed smoothly. The engine ignited, and the rocket lifted up from its pad on Omelek. But 34 seconds into the flight, the engines stopped firing. The team was watching a live feed from a camera, and suddenly the island below the rocket wasn't getting smaller anymore. Someone in the control center said, "That's not good."

Fifty-nine seconds after it lifted off, the rocket crashed into the water just offshore.

Musk had been boasting about this rocket for years, and had invested so much into it. He knew the chances of success were slim, but still, this hurt. All of those hopes vanished in an instant, as pieces of the rocket, once so sturdy and erect, now lay strewn across the reef, polluting the pristine waters like so much litter.

"I remember how sad Elon was that day," said Steve Davis, a young prodigy Musk had hired out of Stanford grad school. "Everyone was just super sad." The rocket had been paid for almost exclusively by Musk, and at some point the money would run out "at which point we're all done," Davis said.

Hans Koenigsmann, one of SpaceX's first employees, was also devastated. "My wife told me that after the first Falcon 1 failed, I didn't talk to her for a month. And I didn't even know. I didn't talk at home at all," he said. "It was a little bit more heartbreaking because we actually collected the wreckage. We picked it up. It's better when it goes out of sight."

SpaceX was trying to make spaceflight affordable, to open it up to the masses. To take humans further than even NASA had. That's what drove them, even then. And if they couldn't get the Falcon 1 to work, it wouldn't just be the failure of a company, it would be as if "we have adversely affected where humanity could go. At least that's how I felt," Davis said. "We all felt that we were kind of the shot to do it. . . . Who else was doing it? There wasn't anyone."

It also hurt because Musk and his team at SpaceX knew that their critics were undoubtedly preparing their I-told-you-so, tsk-tsking postmortems. Legitimate, on-the-level criticism Musk could take; it was the condescension he couldn't stand.

Still, any brooding would have to come later. In the moment, Musk tried to remain cool and professional—the whole team did—as they assessed the situation. The DARPA and air force team was impressed that the people from SpaceX, many of them young and relatively inexperienced, had remained so calm as they watched their rocket hit the water and explode, no matter how badly it hurt inside.

Afterward, Musk tried to stay positive, saying in a statement that "we had a successful liftoff and Falcon made it well clear of the launchpad, but unfortunately the vehicle was lost later in the first stage burn."

They waited for low tide to start to recover the pieces of the rocket. As the SpaceX team pulled pieces of the wreckage out of the water, Weeks, the NASA official on loan to DARPA, tried to impose some order. The debris needed to be cataloged and marked, he said, collected as evidence for the investigation into the cause of the mishap. This needed to be an orderly process, not a free-for-all, yard-sale cleanup.

But Musk, smarting from the sight of his rocket strewn in pieces throughout the reef, was annoyed and pulled Weeks aside. "He wanted to know why I was directing his people," Weeks said. This was his rocket, his explosion, his mess.

Except that it wasn't, not entirely. Weeks reminded him that while it was his rocket, it had just exploded on a government facility on a mission that the government had invested in. There would be a government investigation, and the government would decide when and whether he'd be able to launch again.

"Elon likes to be in control," Weeks said. "And you couldn't blame him. This was his money, and he had put $100 million into it."

DARPA's Mishap Investigation Board, comprising Walker and Weeks and a couple of others working alongside SpaceX, completed its work by July 2006—in less than four months. The problem, it concluded, had been a fuel leak caused by the failure of a single nut of a fuel pump. The corrosive atmosphere of the islands, the salt air that Musk had worried about, had caused it to corrode and then fail.

With the cause found, DARPA cleared SpaceX to get back to flight. Despite the failure, DARPA was ready to stand by SpaceX and fund another try.

The education of SpaceX was only just beginning. And it had learned a powerful lesson about the perils of spaceflight: an entire rocket could go up in flames because of the failure of a single piece of hardware no larger than a pebble.

IN AUGUST 2004, Musk gathered the entire company, all eighty employees, in the cafeteria. He looked somber, or tried to. But he couldn't keep the poker face for long. He was too excited.

"We fucking won!" he said.

SpaceX had won one of the COTS awards, a prize that could be worth as much as $278 million.

The staff cheered wildly. "There was this eruption of joy, people jumping up and down," Shotwell recalled. "This was a huge deal for us, obviously. That it was no longer just Elon pouring his own personal money into it. I think he felt very—I don't think *vindicated* is the right word, but I think he probably felt a great sense of relief that what he had built up until then was worth something, and NASA had recognized it."

After the failure of the Falcon 1 launch, this was an enormous boost, one that gave them hope and motivation to carry on. The other winner was Rocketplane Kistler, an offshoot of the company that had won the sole source contract that Musk had sued over in 2004 and had recently emerged from bankruptcy.

Musk was ebullient, proclaiming it a significant step toward achieving its goal of dramatically lowering the cost of spaceflight. "I think it could be some of the best money NASA's ever spent," he said.

Others weren't so sure they'd be able to drop the cost of space—something companies had been trying to do for decades.

"I'm tired of hearing that; it never comes true," groused John Pike, a space analyst at globalsecurity.org, a think tank. "There have been no improvements of per-pound launch since John Kennedy was president. . . . I think it's just going to be a good way to burn up a bunch of money."

SpaceX and Rocketplane did seem a risky pair for NASA to bet on. Neither company had launched a rocket, and Kistler's finances seemed shaky. It was unclear whether either of them would be able to come close to launching anything safely into orbit—let alone a spacecraft that would have to chase after the International Space Station, orbiting Earth at 17,500 mph. Skeptics inside NASA thought it was nuts, a complete waste of time and money.

To Musk, the supporters "were like the weird rebels within NASA," he said. "They were given a project that everyone expected to fail, and they were just damn determined to make sure it didn't. And they did so because they really believed that things needed to change. They did it for really good reasons. They cared about the advancement of spaceflight, and they really worked their asses off to help us succeed."

But even they admitted they were bushwhacking into unknown territory.

"It was new to everybody," said Marc Timm, a program executive at NASA. "The first meeting, we had no clue how we were going to do it. We had representation from legal, safety, Space Station, NASA Headquarters, and just sat down with a clean whiteboard and said, 'How can we do this?'"

As daunting as it was, it was thrilling to be part of something completely new in an agency that had grown bureaucratic and stodgy, that was often far more comfortable saying no because it was unsafe.

Unlike the cost-plus contracts that sustained the traditional contractors, paying them even when they went over budget or over schedule, these awards were known as Space Act Agreements. The companies would only get paid once they reached certain milestones. By design, the money from NASA was not enough to keep the companies going on their own. They had to find additional outside investment, or customers.

If they didn't, free market forces would take over, and they'd fail.

"Commercial companies had not done this before," said Scott Horowitz, then NASA's associate administrator for the Exploration Systems Mission Directorate. "There was a high risk in them being able to develop the capability, which was supposed to come online when the shuttle retired in 2010. As much risk as there was on the technical side, there was even more risk on the financial side. To be quite honest, the business cases were not that robust."

For years, commercial companies, SpaceX chief among them, had said that they could build rockets more safely and efficiently than NASA. That it was time for the private sector to take over. This was going to be their chance to prove it.

"There's a lot of people that have been running around for years in conferences, yelling and screaming, 'Get out of our way! Give us a chance! We can do it for one tenth of the price, and much more reliably. And a whole lot faster and better than NASA,'" said Griffin, the former NASA administrator.

"They've been screaming that for years. This was like, 'show me.'"

No one was more vociferous than Musk. And he reveled in the opportunity to prove himself, especially with the strict, "show-me" way the payments were structured. The bolder, the better, he figured. The arrangement, he knew, favored a scrappy startup like his, and, for once, put the less nimble, more risk-averse traditional contractors at a strategic disadvantage. They could play it safe. He had nothing to lose.

"The funding is milestone-based, so if we don't achieve milestones, there's no money spent," he said at the time. "It's not like a standard government cost-plus effort where the worse they do, the more money they get. This is the case where if we don't do what we say we're going to do, we don't get paid. So it's a no-lose proposition for the taxpayer."

"I think the fear on the part of some people out there is not that we will fail, but that we will succeed," he added.

For SpaceX, the $278 million was a windfall. The deal was like a "Good Housekeeping seal of approval" that gave the company credibility in the marketplace, recalled Tim Hughes, SpaceX's senior vice president and general counsel. If NASA trusted them, then commercial satellite manufacturers could as well.

But for "a lot of people in the big aerospace world this was considered a token amount of money to placate these commercial guys so they could stop complaining to Congress," Musk said, years later. "Just give them enough to hang themselves. . . . Let's just give these annoying commercial people enough money so that they can fail, and then we can say that was dumb. We don't have to do that again."

SpaceX was still not regarded as a threat. Not by the Lockheeds and Boeings, which had large, multiyear contracts in hand that could sustain them as long as there was support in Congress. And while the program was a bold step for NASA, one that SpaceX and other startups jumped at, it was largely dismissed by the rest of the industry. Lockheed Martin, Boeing, Northrop Grumman all passed, not thinking that this new era would last.

They were focused on another NASA program that was a central part of the Bush administration's official "Vision for Space Exploration." Called Constellation, it was the White House's grand plan to bring humans back to the moon by 2020 and eventually to Mars. Instead of just a few hundred million dollars at stake, the construction of a pair of new rockets, the Ares I and V, a lunar lander, and a spacecraft, the Orion, was a prize worth tens of billions over many years. In comparison, the COTS program was mere crumbs. COTS was a side program, carved out by Griffin; Constellation was the big program of record that the big contractors remained focused on.

"They screwed themselves because they were just arrogant and complacent," Musk said later. "Look, Boeing doesn't get out of bed for less than a billion."

(In response, Boeing noted that "Apollo and the programs that inspire all space enthusiasts would not have been possible without Boeing. At the turn of the 21st Century, before Musk entered the space business, Boeing was building the International Space Station with NASA, where we've kept astronauts safe and continuously on orbit for over seventeen years." And the company said that "while others talk about aspirations and hopes, we actually do things in space and will deliver on our commitment to America's journey to Mars. That's what we get out of bed for.")

Looking back on it, Griffin said he was surprised that the legacy contractors "did not see the handwriting on the wall that there would be in the future some commercial space enterprises." They all had the "muscle mass," he said. They were the "trained athletes on the field," who could have "undercut even the best entrepreneurial firm" had they chosen to bid on COTS. Had they wanted it, the prize was theirs for the taking.

Griffin saw a lesson that could be taught in business school. NASA, the customer, was indicating it was moving in a different direction. But its traditional contractors ignored it, or didn't take it seriously, or weren't nimble enough to move with it. That created an opening. A small one. A mere crack a couple million dollars wide that was easy to ignore when they were marching through a gilded, billion-dollar doorway.

"The entrepreneurial firms can exist only if the larger established contractors create a niche for them," Griffin said. "If I'd been running Boeing or Lockheed or Northrop, I would not have done that. I would not have allowed a fledgling competitor to invade my space."

TWO YEARS LATER, however, the invaders were stumbling. The insurgency that COTS was supposed to foment was flaming out in SpaceX's failed launch attempt, and subsequent delays, which fueled doubt and a lack of funding that put Rocketplane Kistler out of business. The millions that NASA had awarded the company as part of the COTS program were not enough to sustain it. And now the space agency was forced to go out and find another company to participate in the program, as critics questioned whether these young, inexperienced companies were worth NASA's time and investment.

After its 2006 failure, it took SpaceX a year to fly again. This time, on March 20, 2007, it performed much better, flying to 180 miles, well past the edge of space. Its second stage separated, and the onboard camera showed the booster falling back toward the ocean below. In the control room, the SpaceX employees could see the curvature of Earth and the blackness of space.

"I'm going to watch that video for a long fucking time!" Musk said. "Congratulations, guys."

But then, before the second stage could get to orbit, the second stage started wobbling uncontrollably before it fell back to Earth.

"This was a pretty nerve-racking day," Musk said afterward. "The rocket business is definitely not a low-stress business, that's for sure."

Still, he found solace in the fact that the rocket had indeed made it to space. "I don't think I'm disappointed," he added. "In fact, I'm pretty happy."

As he later pointed out, it was a test flight. The whole point was to see how the system worked, and root out any problems.

On August 3, 2008, the third launch attempt also came just short of getting to orbit. At stage separation, the first and second stages collided, causing yet another failure.

Publicly, Musk remained resolute. "SpaceX will not skip a beat in execution going forward," he said. "We are in a very good financial basis here. We have the resolve. We have the financial base. And we have the expertise."

He added: "For my part, I will never give up, and I mean never."

Despite those rosy predictions, the truth was, SpaceX was hurting. Musk was burning through the $100 million he had invested of his own money, while it had also earned a $20 million investment from Founders Fund, the Silicon Valley venture capital firm started by Peter Thiel, who knew Musk from their early days at PayPal..

The challenge "was really keeping the company going financially while we were struggling. That was my contribution primarily," SpaceX president Gwynne Shotwell recalled. "I didn't get to do as much engineering as I would have liked to, but continually convincing customers to invest in SpaceX, and to take the risk associated with buying launches from us. I was focused on keeping the company alive, keeping people paid while we were struggling and getting through it. Because I knew we'd get through it. Technically, I knew we would get through it. It was a matter of could we get through it financially and stay stable."

The three failures were "really incredibly painful," Musk said later.

If this fourth launch didn't work, "we would have ceased to exist. I was out of money."

LESS THAN TWO months later, on September 28, 2008, SpaceX tried again. To prevent the first and second stages from colliding, the engineers adjusted the timing ever so slightly.

"Between the third and the fourth flight, we changed one number, nothing else," said Hans Koenigsmann, SpaceX's vice president of mission assurance. "That was the time we needed to separate the two stages."

That was the technical solution. Tim Hughes, SpaceX's general counsel, had a more superstitious one: adding a pair of four-leaf clovers to the mission patch. Throughout NASA's history, the agency had created patches for each launch, artistic talismans that captured the spirit of the adventure while, hopefully, bringing luck. It was a ritual that dated back to Mercury, Gemini, and Apollo, and on through the shuttle. Astronauts were like ballplayers, believing that the power of symbols and a certain set of rules, however absurd—never step on the foul line—bequeathed success, a line drive up the middle, or not dying in a fireball just off the launchpad.

Knowing the importance that symbolism and superstition had for its astronauts, NASA allowed its crews to design their own patches. For Apollo 11, which would become one of the most recognizable, the astronauts decided to omit their names in a self-effacing gesture, though listing the crew members had been the practice in the past.

"We wanted to keep our three names off it because we wanted the design to be representative of everyone who had worked toward a lunar landing," said Michael Collins, a member of the Apollo 11 crew. "And there were thousands who could take a proprietary interest in it, yet who would never see their names woven into the fabric of a patch. Further, we wanted the design to be symbolic rather than explicit."

Neil Armstrong didn't want the word *eleven* spelled out, as it was in the original design, because non-English speakers wouldn't be able to read it. So, instead it would go by the numeral, 11.

SpaceX, now desperate for what could be its fourth and final flight, was willing to believe wholeheartedly in the power of the patch. It needed all the luck it could get. The four-leaf clovers were added to the patch.

THIS TIME THE Falcon 1 rocket got to orbit in a flawless flight that was perfect from the countdown to stage separation, which ended with a crescendo of cheers from SpaceX's more than five hundred employees.

The company had broadcast the launch on the web, in a flawed, rudimentary real-time show anchored by amateurs whose enthusiasm and passion were genuine—they were SpaceX employees, after all—but who were ill trained for the rigors and length of almost an hour on air.

The moment Falcon 1 had achieved orbit, the commentator declared that "Falcon 1 made history as the first privately developed launch vehicle to reach Earth orbit from the ground."

He then took a moment to explain the long and improbable odyssey that had led to this feat:

"SpaceX has designed and developed this vehicle from the ground up, from a blank sheet of paper they've done all the design, all the testing in house. We don't outsource, and we have achieved this with a company that is only now five hundred people. And it has all occurred in under six years."

For this launch, Musk was in California with the bulk of his team, not in Kwajalein. He emerged before the throng in a nondescript polo shirt, trying to rise to the occasion for his cheering employees standing around him three and four deep. Some had brought their children to the factory to witness the launch, and they hoisted them on their shoulders for a better view of Musk. But he was utterly overwhelmed and at a loss for words.

"That was freakin' awesome," he managed in a halting cadence, then stated the obvious: "We made orbit."

After thanking his team, he said, "There were a lot of people who thought we couldn't do it—a lot actually." He let out a little maniacal, take-that laugh of redemption, before continuing.

"But as the saying goes the fourth time's the charm, right? This really means a lot to SpaceX. Getting to orbit, that's just a huge milestone. There are only a few countries that have done it. It's normally a country thing, not a company thing. It's an amazing achievement."

As the words left his mouth, it was as if just then, for the first time, he realized the enormity of the accomplishment—*it was what countries had done*. And as that notion settled in, he once again became speechless.

"My mind is kind of frazzled," he stuttered, exhausted. "Man, it's definitely one of the greatest days of my life, and probably for most people here. We've shown people we can do it. This is just the first step of many."

The Falcon 1 would lead to the Falcon 9, an even more powerful rocket—with nine engines, compared to one—that was under development, he said. And then there was the Dragon spacecraft, a capsule being designed to take cargo to the International Space Station. That was the next prize, the natural follow-up to the COTS program, which, after all, was supposed to help develop the capability to at least fly supplies—food, equipment, science experiments, toilet paper—to the station.

It was as if the opening of this one door would lead to another and then another, achievement begetting monumental achievement, one small step, leading to one giant leap springing outward to a point on the horizon that only he could see.

What else was possible?

"A lot of things," he said, allowing his mind to wander into the future.

"Ultimately," he said, "I think even getting to Mars."

That remained to be seen. But from then on, every single SpaceX mission patch would have a common feature, one often hidden in the recesses: a four-leaf clover.

AT THE END of 2008, SpaceX got a gift two days before Christmas: a $1.6 billion contract from NASA to fly cargo in its Dragon spacecraft on as many as twelve flights to the International Space Station. For SpaceX, it meant that the company had finally arrived, and had been given NASA's full, unmitigated endorsement. These would be real missions—delivering thousands of pounds of supplies to the space station—feats that would require the company to successfully launch its Dragon capsule to orbit, which would then fly and dock with the space station, traveling at 17,500 mph. It was an achievement that had been accomplished only by nations, and only a select few—the United States, Russia, Japan, and the European Union.

When Musk got the call from NASA, he was overcome. The last few years had been a struggle, filled with failed launches, even a divorce. But now six years after founding his space company, he had been vindicated. And when the NASA officials finished telling him that SpaceX had won the contract, he unceremoniously blurted out, "I love you guys!"

He also changed his login password to "ilovenasa."

WHILE MUSK WAS charging ahead, Bezos was still moving slowly, taking small, deliberate baby steps. More than a year after Blue Origin had flown Charon, the massive dronelike vehicle powered by the four jet engines acquired from the South African Air Force, the company was ready to fly again, this time with something that looked like a big, white gumdrop, a flying, flat-bottomed Humpty Dumpty.

Before dawn on November 3, 2006, a cold morning in West Texas, a flatbed truck rolled the vehicle out of its hangar to the launchpad. And as the sun came up, turning the distant mountains purple, employees and their families gathered in the bleachers, some with blue cheerleader pom-poms, ready to watch the launch on a jumbotron.

The vehicle they had come to see was named Goddard, after Robert Goddard, the father of modern rocketry, who in 1926 became the first to ever launch a liquid-fueled rocket. He was a builder and a dreamer, who in 1919 wrote a paper, "A Method of Reaching Extreme Altitudes," published by the Smithsonian Institution, that touched on the possibility of developing a rocket to reach the moon.

At the time, the notion of reaching the moon seemed as far-fetched as it was ridiculous. Goddard was derided as "moony" and "crackpot," and even the *New York Times* wrote a scathing editorial in 1920 under the headline "A Severe Strain on Credulity," which scoffed at the idea, saying a rocket could not work in the vacuum of space.

"That Professor Goddard, with his 'chair' in Clark College and the countenancing of the Smithsonian Institution does not know the relation of action to reaction, and of the need to have something better than a vacuum against which to react—to say that would be absurd," declared the *Times*. "Of course he only seems to lack the knowledge ladled out daily in high schools."

Goddard responded by saying that "every vision is a joke until the first man accomplishes it; once realized it becomes commonplace."

But the ridicule led Goddard, a shy man who preferred working alone, to an even greater measure of reclusiveness and dedication to a long-term vision of spaceflight that he knew would take many decades to fulfill.

"How many more years I shall be able to work on the problem I do not know," he wrote in 1932. "I hope as long as I live. There can be no thought of finishing, for aiming at the stars, both literally and figuratively, is the work of generations, so that no matter how much progress one makes, there is always the thrill of just beginning."

He died in 1945, without having lived to see humans go to space. But just before the Apollo 11 moon landing in 1969, he received a belated, postmortem vindication. By then, it was abundantly clear that rockets could indeed work in space, and the *Times* issued a correction to its editorial, a half-century after it was published.

"Further investigation and experimentation have confirmed the findings of Isaac Newton in the 17th century, and it is now definitely established that a rocket can function in a vacuum as well as in an atmosphere," it read. "The Times regrets the error."

FROM A PURELY chronological standpoint, it made sense that Bezos named his first rocket after the father of rocketry. But they were also kindred spirits. Like Goddard, Bezos was dedicated to taking the long view, with Blue seeing it as an enterprise that would take generations to complete. Like Goddard, Bezos believed that the impossible could be made routine. And like Goddard, Bezos's company shunned the press, keeping its work secret, carefully protected from scrutiny and the criticism that would surely follow.

In fact, Bezos was such a fan that he chose "Goddard" as the middle name for one of his sons.

After years of work, Bezos and Blue Origin had arrived at the debut of its first launch, modest though it would be. Still, there was indeed a feeling of the "thrill of just beginning," and the company made it festive. A cowboy cooked biscuits over an open fire. For the kids, there was a bouncy castle.

The countdown played over a loudspeaker. Goddard lifted off, flying to 285 feet and then touching back down in a flight that lasted all of thirty seconds. Bezos celebrated with a giant bottle of champagne, and joked that his only job during the launch was to pop the cork. The launch was more successful than the uncorking, though. Bezos broke off the top of the cork, leaving the rest still in the bottle.

Goddard's short hop was another small and modest step forward for the company, and it stood in stark contrast to the giant leap Musk had taken. SpaceX's first successful launch wasn't just a couple hundred feet. It wasn't even a suborbital spaceflight. SpaceX instead went straight to orbit, achieving an extraordinarily difficult task that requires the rocket to go so fast that it essentially is continually falling around Earth. It was typical SpaceX: *Head down. Plow through the line.*

SpaceX was now looking ahead to its more powerful Falcon 9 rocket, which would launch from Pad 40 at Cape Canaveral Air Force Station. But Blue Origin remained steadfastly committed to its approach, even if it seemed the company was barely leaving the gate.

In following this deliberative path, Goddard would be followed by New Shepard, a rocket named for Alan Shepard, the first American to reach space. The company's progress, then, would mimic the step-by-step evolution of American spaceflight. Goddard's rocket design helped lead to Shepard's suborbital, up-and-down flight in 1961, which lasted just 15 minutes and 28 seconds. NASA didn't put a man into orbit until John Glenn circled Earth the following year.

Blue Origin didn't say anything publicly about its launch or its next steps, until two months later. In a blog post, Bezos wrote:

"Accomplishing this mission will take a long time, and we're working on it methodically. We believe in incremental improvement and in keeping investments at a pace that's sustainable. Slow and steady is the way to achieve results, and we do not kid ourselves into thinking this will get easier as we go along. Smaller, more frequent steps drive a faster rate of learning, help us maintain focus, and give each of us an opportunity to see our latest work fly sooner."

It didn't matter how far ahead the hare was; the tortoise was content to keep the methodical pace Bezos had outlined for the company in his 2004 letter: "Be the tortoise and not the hare." It would stay true to its motto, "Gradatim Ferociter," or "Step by Step, Ferociously," while repeating over and over:

Slow is smooth and smooth is fast. Slow is smooth and smooth is fast. Slow is smooth and smooth is fast.

9

"Dependable or a Little Nuts?"

I T WAS JUST lying there, like a piece of used furniture left out on the curb for anyone to take. A massive 125,000-gallon liquid nitrogen tank as big as one of those water storage bubbles with the name of the town it supplies painted across it. A SpaceX employee just happened to spot it while driving past an abandoned launchpad at the Cape Canaveral Air Force Station and thought, *Maybe we could use this?*

The company had signed a lease for its own launch complex there, Pad 40, which since the 1960s had been used for the military's Titan rockets. SpaceX had just knocked down the old structure that had been there for years. Now, in 2008, SpaceX was rebuilding the facility—on the cheap—for its new Falcon 9 rocket and Dragon spacecraft to deliver cargo to the International Space Station.

Despite having sat outside for years, the liquid nitrogen tank seemed in decent shape, and Brian Mosdell, who was leading the small team of ten SpaceX employees tasked with rebuilding the launchpad at the Cape, wanted it.

They called the US Air Force again and again, seeking permission to take it. But no one there could be bothered; to the air force, the thing was a hunk of junk, not worth wasting any time on. Finally, Mosdell's crew got a response, and was put in touch with a company that had been hired to haul away the tank and destroy it. The company was willing to part with it for $1 over the cost of scrapping it—$86,000.

Mosdell bought it, and then spent about a quarter of a million dollars refurbishing it. Even with that, the cost was far lower than building a new tank from scratch, which he estimated would have cost more than $2 million. Musk was thrilled with the team's ingenuity, showing off the tank on a video tour of the pad. "Here we are on top of our giant ball of liquid oxygen," he said. "They say SpaceX has big balls, and it's true."

An engineer, Mosdell had spent twenty years working at various launch facilities on the Cape, serving in a variety of roles with major defense contractors, including General Dynamics and Boeing/McDonnell Douglas. But when SpaceX hired him in 2008 after he had been working at its rival, the United Launch Alliance, he quickly realized this startup of a space company was unlike any other. When he was at Boeing years before, he'd actually tried to salvage the used liquid nitrogen tank.

"But everybody shot that down," he said. "Nobody was interested. It was too difficult." His bosses wanted to know, "Who would we even call to get it?"

When he worked for large defense contractors, "there was never a mind-set or interest in reusing things," he said. "Everything needed to be built from the ground up. It's government contracts and government money."

The rules were the rules, and the price was the price. No one questioned the cost, or the regulations, or the system. That's how it was done.

Until SpaceX. It had an altogether different mind-set, an obsession with finding ways to do things cheaply and efficiently, and an almost instinctive contrarian bent that questioned everything—the price, the rules, the old way of doing things. If Cape Canaveral and its leaders were the adults, SpaceX was the child, constantly curious, always asking why.

When Mosdell got a call "out of the blue" from SpaceX, he wasn't sure what to think. He had a safe, comfortable job at the Alliance, and like many there, he didn't think SpaceX was a serious player. "I thought they had a PowerPoint and paper rocket," or one that just existed in theory, he said. His colleagues at the Alliance wondered why he'd want to go work at a company like SpaceX, which had accomplished relatively little.

"Like me, they didn't see any threat coming out of SpaceX," he said.

Or any future.

But when he went out to California for his interview, "everything really changed my mind," Mosdell recalled. "I saw at least $25 million of flight hardware in various stages of fabrication. That's when my head snapped and I said, 'Hey, wait a minute. This is the real deal.'"

In interview after interview, the executives stressed that SpaceX was different than any of the companies listed on his resume. "This is not heritage aerospace," he was told. "We're lean and mean. If you come work for us, you're going to have a lot of creative license. You're not going to get stifled by bureaucracy."

The corporate culture was freewheeling and hard charging, with a mix of industry veterans and young kids with virtually no experience in building rockets, but who were brilliant and willing to devote themselves to the cause with abandon.

It wasn't a place for everyone. The whole enterprise of building rockets seemed a little crazy. The hours were ridiculously long; the work, challenging. It was great for young, energetic, and brilliant workaholics, but not so great for those seeking a "work-life balance." Elon Musk was demanding and known to yell at employees on the middle of the factory floor. A former executive at Lockheed who got to know Musk and the culture at SpaceX couldn't believe how relentless and demanding they were. "If I did that, at a public company, the HR person and the lawyers would be in my office within ten minutes to ship me off to eighteen months of sensitivity training," he said.

Hans Koenigsmann, SpaceX's vice president of mission assurance, praised Musk for the alchemy he performed at the company: "The way Elon turns the future into reality is pretty amazing," he said. "The whole enterprise is pretty much against the odds." But he also didn't think he'd grow old at the company; he just didn't have the energy. "It takes a toll on you, and it's hard to do a half job at SpaceX."

Musk was aware of how demanding he and SpaceX could be. "Some of the guys kind of got burned out," he said. "They just got fried after too much intensity."

Musk hired very smart people who had to prove their proficiency in personal interviews with him. Engineers stood atop the corporate totem pole, with everyone else behind. "SpaceX had what Elon called a high signal-to-noise ratio, meaning that people who added value were engineers. They were signal," said Tim Hughes, the company's general counsel. "And people who were nonengineers for the most part were noise."

One of the early hires was Mark Juncosa, who came to the company right out of graduate school, lured by Musk's smarts and passion and the whole wild vibe of the company, which included Musk's edict allowing employees to get up and leave meetings they didn't need to attend. No questions asked.

"There were a lot of people that were quite bright," said Juncosa, who would become the company's vice president of vehicle engineering. "And not boring. They all had a big fire under their ass and were quite crazy."

Juncosa wasn't sure that the company would ever be successful. "How are we going to figure out how to make a spaceship that took an incredible amount of people in the sixties, when we didn't have those resources?" he wondered.

But here they were "fighting our fucking hardest," all following Musk, believing that he always "figures out how to make the magic happen."

BY THE TIME Mosdell showed up, the company had moved into a new, bigger facility in Hawthorne, a former Boeing 747 fuselage factory not far from the Los Angeles airport. For an aerospace engineer, it was like Willy Wonka's chocolate factory. Massive rockets were being built from scratch, long cylindrical cores stretching along the factory floor like the hulls of great ships. Engines—new, American-made engines—were being manufactured in-house. And hundreds of workers, many of them so young it seemed as if they should still be in college, fanned out across a floor humming with what Mosdell saw as a sense of let's-get-it-done urgency.

The company often warned job applicants that their interview with Musk could be short and awkward because he might be multitasking through it, or take long pauses to think during which he said nothing for minutes on end. Mosdell found Musk a touch awkward and abrupt, but smart. Mosdell had showed up prepared to talk about his experience building launchpads, which, after all, was what SpaceX wanted him to do. But instead, Musk wanted to talk hard-core rocketry. Specifically the Delta IV rocket and its RS-68 engines, which Mosdell had some experience with when at Boeing.

Over the course of the interview, they discussed "labyrinth purges" and "pump shaft seal design" and "the science behind using helium as opposed to nitrogen." Mosdell didn't know whether Musk was testing his knowledge or genuinely curious. And then it was over.

"He abruptly said, 'Okay, great, thanks for coming in.' And spun his chair around and went back to his computer," Mosdell said. "I couldn't tell if it went well or not."

When he was hired, Mosdell became SpaceX's tenth employee at the Cape, and was put to work almost immediately, rebuilding Pad 40. Eager to show he could be resourceful, Mosdell and his team became the scavengers of Cape Canaveral, going around looking for leftover hardware as if they were on a treasure hunt.

So, the old railcars from the 1960s that had once been used to ferry helium between New Orleans and Cape Canaveral became new storage tanks. "We took off the wheels and basically set them up on fixed pedestals," Mosdell said.

Instead of spending $75,000 on new air-conditioning chillers for the ground equipment building, someone found a deal on eBay for $10,000.

In addition to recycling old material, they pushed back on regulations they saw as anachronistic leftovers from an earlier era.

When the company was told it would cost $2 million for a pair of cranes to lift the Falcon 9, for example, it questioned the price, wanting to know why it was so expensive. The reason was that the air force required the cranes to meet a series of safety requirements to prevent, say, a hook from suddenly dropping too fast. But modern technology had rendered many of those requirements, some decades old, unnecessary.

Mosdell and the SpaceX team lobbied the air force officials at Cape Canaveral, ultimately convincing them to strip out many of the old regulations that were driving up the price. They did, and SpaceX was able to purchase the cranes for $300,000.

Then, the air force said that Pad 40's flame duct needed to be extended with a water system. Bids for a traditional concrete trench to guide the rocket flames out and away from the pad came in at about $3 million. Mosdell thought they could do better.

"Ultimately, the engineering team designed, and the pad crew built, a flame duct extension using steel box beams, which also carried cooling and acoustic suppression water inside the beams," he said.

The result: a system that met the air force's requirements, for a tenth of the cost.

"We had to be superscrappy," Musk said. "If we did it the standard way, we would have run out of money. For many years we were week to week on cash flow, within weeks of running out of money. It definitely creates a mind-set of smart spending. Be scrappy or die: those were our two options. Buy scrap components, fix them up, and make them work."

Cost drove lots of decisions, even how the company would build its rockets. Although some companies assembled their rockets vertically on the pad, that required what was known as a mobile service tower. They were giant structures that would surround the rocket while it was being built and then get wheeled away.

"Elon is, like, 'That is the dumbest thing I've ever heard of ever,'" Gwynne Shotwell recalled. "'Like, how expensive and inefficient is that?'"

SpaceX built its rockets in its California factory "where it's all clean and neat and nice anyhow," said Shotwell. By building them horizontally, you reduce the risk of having employees as they work high off the ground, she explained.

When it was building Falcon 1, the company bought a theodolite, a tool used to align the rocket, on eBay, saving the company $25,000.

In the same frugal manner, the Dragon spacecraft looks the way it does because it's the simplest design.

"If someone says design a reentry capsule, and you give it to NASA or someone else, they are going to spend, like, a year designing its shape," said Steve Davis, SpaceX's director of advanced projects. "For us it was, the bottom is the diameter of the Falcon 9. Because it was on the Falcon 9. The top is the diameter of the [port where it would dock with the space station]. The design is now complete. That was it. Connect two lines."

The rocket's avionics were powered by a $5,000 computer instead of the much more expensive aerospace hardware. One employee even found a piece of metal in a junkyard that he thought might be used as part of the rocket's fairing, the protective cone on the top of the rocket that shields the payload, such as a satellite.

Instead of using the straps that came with the module used for storing cargo, which NASA's astronauts on the space station apparently found cumbersome, they found straps using a NASCAR design that the astronauts loved.

SpaceX even questioned the kinds of latches that were used in the lockers of the space station. Each locker required two latches, which each cost $1,500 and had twenty to twenty-five parts. At "SpaceX, we weren't going to build that," recalled John Couluris, of SpaceX mission operations. "One engineer was inspired—I think it was honestly in the men's room—where he saw the latch on a stall, and we were able to make a locking mechanism out of that."

Instead of $1,500, it cost $30. "It's more reliable, and it's easier to replace if it ever goes bad," Couluris said. "The astronauts, not only did they love it, but they loved the story behind it because that shows the ingenuity."

Once Musk got wind that the air-conditioning system used to keep the satellite cool in the rocket's fairing, or nose cone, was going to cost some $3 million or $4 million, he confronted the designer in his cubicle about it.

"What's the volume in the fairing?" he wanted to know. The answer was a couple thousand square feet, less than the size of a house.

He turned to Shotwell and asked her how much a new air-conditioning system for a house cost.

"We just changed our air-conditioning," she replied. "It was six thousand bucks."

"Why is this three or four million dollars when your air-conditioning system is six thousand dollars?" he demanded again. "Go back and figure this out."

They did: they bought six commercial A/C units with larger pumps that could handle a larger airflow.

It was new and innovative and totally different than anything NASA's leadership was used to. And it took some convincing that, though different, SpaceX's approaches were still okay.

"The biggest challenge, I think, that we had in the execution of this was convincing NASA, every step of the way, that though we're going to do business very differently, we're going to get it right," Shotwell said. "Because no one had experience doing this job the way we wanted to do this job. Unfortunately, the industry is frankly, I think, hampered—I've been doing this for thirty years, so I think I can say that—by cost-plus contracts. The incentive on a cost-plus contract is not to minimize cost, it's to maximize effort. Our philosophy was not minimize effort, but op-timize effort."

NASA officials were, at first, flabbergasted. But SpaceX ultimately won them over with its scrappy, Silicon Valley ethos.

"When we talked to them about a given part or component in the de-sign, they'd say, 'Well, we could go buy this from this vendor, but it's like $50,000. It's way too expensive. It's ridiculous. We could build this for $2,000 in our shop.' Almost every decision that they made had cost built into it," said Michael Horkachuck, a NASA official, who worked closely with SpaceX on the Commercial Orbital Transportation Services (COTS) program.

"It was unique because I almost never heard NASA engineers talking about cost of a part when they were making design trades and decisions. They were worried about is it going to work, is it going to work reliably and safely and meet all the requirements? Cost generally, although a fac-tor, was never really in the forefront as much as mission success.

"Here they were making more trades on, 'Well, you could do it that way, but this way is a whole lot cheaper and probably just as good.' That was a different mind-set, and I think something that maybe the rest of the agency needs to look at a little bit more."

In 2008, NASA was building a big rocket, too. A pair, actually—the Ares I, which would fly to low Earth orbit, and the Ares V, which was destined for the moon and then Mars. Along with the Orion spacecraft, their names, taken from Greek and Roman mythology, matched the lofty aims of what was then a White House program called Constellation.

They were part of President George W. Bush's plan—called the Vi-sion for Space Exploration—for the nation to return to the moon. In a speech at NASA headquarters attended by Eugene Cernan, the last man to walk on the lunar surface, the president recited what Cernan had said as he departed the moon, promising that "we shall return." In his speech, Bush promised that "America will make those words come true."

But in 2008, when Barack Obama was elected president, his NASA transition team led by Lori Garver, a space agency veteran who had also advised Hillary Clinton, had promised to "look under the hood" at Con-stellation. And when it did, it found all sorts of problems. The costs were skyrocketing, the schedule was slipping, and there was frustration that yet another attempt by a president to re-create the Apollo magic was falling short.

Bush's plan became fodder for late-night television, which mocked an ambition for space exploration that a generation earlier had been venerated for achieving the impossible. Not that long ago, the United States had reached the moon, but its space program had since had so many false starts, been subject to so many unfulfilled political promises, that now the critics were quick to pierce the soaring rhetoric and bring it back to the ground.

"He wants to build like a space station on the moon, and then from the moon, he wants to launch people to Mars," David Letterman said in one of his monologues. "You know what this means, ladies and gentlemen? He's been drinking again."

Cernan's prophecy was in trouble. And his prediction that Apollo 17, the last of the lunar missions, "was the end of the beginning, and not the end," seemed increasingly hollow. By 2008, when Obama was elected, the moon seemed as distant as ever. The Next Giant Leap was going to once again be postponed.

THE OBAMA WHITE House hadn't yet gotten to nominating a new NASA administrator but was already moving ahead to manage what it saw as a crisis. The troubled space shuttle program was scheduled to retire in 2010, and with Constellation running years behind schedule, that meant NASA would lose the ability to fly astronauts from US soil.

After fifty years of historic launches—from Alan Shepard's suborbital voyage and John Glenn's orbit to the Apollo moon missions and the shuttle—the United States would have to rely on Russia—the country it had bested in the race to the moon—for rides to the cosmos.

On December 8, 2008, one month after Election Day, a "private and confidential" forty-five-page memo written by President Obama's NASA transition team identified the gap in flight from US soil as the "biggest near-term challenge for NASA."

"The Ares I/Orion programs are budgeted at close to $15 billion over the next five years, have significant technical challenges, and are at least two years behind schedule," it read. The new NASA leadership would have a choice to make about the future of the program—and of NASA.

One option was to rely even more on the commercial sector. If such companies as SpaceX and Orbital Sciences could fly cargo to the space station, then they could also be trusted to fly astronauts there. And maybe they could do it even cheaper and faster.

"While investment in new technology is inherently risky, funding a commercial crew transport effort. . . . may be a viable alternative or complement to Ares I/Orion," the memo continued, "and would support President-Elect Obama's commitment to drive economic and technological innovation through public/private partnerships."

To many on the transition team, the Constellation program was a holdover not just from a previous administration, but from an old way of thinking about how to get to space—big government programs that satisfied Congress because they led to jobs. But they rarely, if ever, produced reliable and efficient transportation to space.

As the new leadership came into NASA after Obama was inaugurated in 2009, the frustration with NASA's inability to advance further since Apollo was felt acutely. One member of the new leadership team circulated a PowerPoint slide laying out the retreat in American human spaceflight, as compared to commercial aviation:

In 1961, [Yuri] Gagarin was first person to fly in space
—48 years later (Year 2009)
  • 46 people expected to travel to low Earth obit
  • Risk of loss of life estimated at 1 in 254

In 1903, Wright Brothers flew at Kitty Hawk
—48 years later (Year 1951)
  • 39 million passengers flew on commercial airlines
  • 1 fatal accident in 288,444 departures

The "lack of progress over 50 years is partly caused by [a] closed approach innovation," the slides stated.

The new leadership's solution would be an "open innovation strategy to stimulate commercial spaceflight capability."

Nearly a decade after Andy Beal folded his company, complaining that he couldn't enter the closed market, here NASA was acknowledging that it needed to open up.

One of the most significant ways to do that would be to start what it called a "commercial crew" program, a competition to fly NASA astronauts to the space station.

By 2009, it was becoming increasingly clear there was the political will inside the White House to kill Constellation and take on the massive political fight in Washington that would ensue. Constellation's big, legacy contractors were well established and well connected. But an internal memo at NASA took aim at Bush's Vision for Space Exploration, with a title that read "Canceling Constellation: Bringing Reality to the Vision."

The paper, marked "for NASA internal use only," acknowledged that the "vision has had widespread support, but that vision is far different from the realities of the actual Constellation program."

The realities were that the program "is not on track to return humans to the moon by 2020, as is widely believed." The program "eschews technology development." The cost of the Ares I rocket had "tripled in four years," whereas the cost of the Orion spacecraft "doubled in four years."

But killing the program would be tough. NASA had already awarded more than $10 billion for Constellation. And while a lobbyist from SpaceX commended NASA's new leaders for the bold move, he warned them in an e-mail that it would be rough:

"The only thing more difficult than killing a big Lockheed or Boeing government program is killing a Lockheed AND Boeing program," he wrote.

On November 16, 2009, White House budget director Peter Orszag and John Holdren, the president's assistant for science and technology, wrote a memo to the president, saying that Constellation "is over budget, behind schedule, off course, and 'unexecutable.'"

That last word—*unexecutable*—had been the conclusion of an independent commission, led by Norman Augustine, the former chief executive of Lockheed Martin. If Augustine, the ultimate industry insider, could be so critical of a program that his former company was involved in, the Obama administration had the cover it needed. And it began to move to cancel Constellation.

"The estimated cost for developing the Ares I and Orion has increased from about $18 billion to $34 billion," the two senior members of the White House wrote to the president. "Once the vehicles are available, operations costs are also expected to be high, ranging from $2 billion to $3.6

billion a year to service the Space Station. (Russian spacecraft cost $300 million to $400 million per year to conduct these missions.)"

Obama highlighted the last two sentences, and scribbled a handwritten note in the margin: "What explains the vast difference?"

The difference was that the United States was paying the Russians for just a taxi service to the station—and not for the development costs for the rockets. But still, the numbers were glaring.

And they now had the attention of the president.

IN EARLY 2010, Obama pulled the trigger, canceling Constellation outright. Just as the SpaceX lobbyist had predicted, an intense fight broke out in Washington.

"The president's proposed NASA budget begins the death march for the future of U.S. human spaceflight," said Senator Richard Shelby, the senior Republican member of the appropriations committee. "If this budget is enacted, NASA will no longer be an agency of innovation and hard science. It will be the agency of pipe dreams and fairy tales."

Michael Griffin, the former NASA administrator who had started the Constellation program, said canceling it "means that essentially the U.S. has decided that they're not going to be a significant player in human space flight for the foreseeable future. The path that they're on with this budget is a path that can't work."

Obama's plan to rely on the commercial sector to ferry astronauts to the station as part of the commercial crew program also faced immediate skepticism.

"One day it will be like commercial airline travel, just not yet," Griffin said. "It's like 1920. Lindbergh hasn't flown the Atlantic, and they're trying to sell 747s to Pan Am."

The White House could perhaps withstand the criticisms of the former NASA administrator. It could maybe even weather the attacks from Shelby and his cronies in the Senate. But soon it had an even bigger problem on its hands—one it didn't foresee: Neil Armstrong and some of his fellow astronauts, who wrote a scathing letter to Obama decrying the decision to cancel Constellation along with the shuttle program.

The founding fathers of space were furious, and saw the decision as an abdication of the dreams of their generation, who at the "because it is hard" urging of John F. Kennedy had pulled off the moon landing and had expected that the first small steps of the Space Age would be followed by the promise of giant leaps for humanity, to Mars and beyond. But now the astronauts saw only retreat, and troubling signs that suggested the Apollo moon landing was a fluke, a onetime feat never to be repeated.

After Apollo, president after president had promised the next great American adventure in space, new lunar missions, even Mars. But years passed, then decades, and NASA remained confined to the low Earth orbit of the International Space Station, a mere 240 miles away. It was as if Columbus had discovered the New World and no one followed.

THE WHITE HOUSE was losing the public relations battle. Neil Armstrong, of all people, had come out against the decision. For the Obama administration, the timing could not have been worse. It was in the middle of the fight to push through the health-care initiative, the Affordable Care Act, popularly known as Obamacare, and it didn't need any distractions. Particularly not one in space, which wasn't a high priority. And it certainly did not need America's heroes criticizing the nation's chief executive.

This needed to be fixed, and fast. So the White House decided to bring out the president himself to help remedy what was turning into an increasingly difficult problem. Aides started preparing a speech to be delivered on April 15, 2010, at the Kennedy Space Center, the first and only major space address of Obama's presidency, one that would show that the United States would retain its leadership role in space.

The president began by acknowledging Buzz Aldrin, who supported the decision to ditch Constellation and whose seat front and center was a not-so-subtle counter to Armstrong, Cernan, and Lovell. And then Obama took aim at Bush's plan to return to the moon.

"I just have to say pretty bluntly here: we've been there before," he said. "Buzz has been there. There's a lot more space to explore, and a lot more to learn when we do."

The nation wouldn't abandon plans for deep space exploration, he promised. Astronauts would, for the first time, land on an asteroid. By the 2030s, a crew would fly around Mars, he said, "with a landing on

Mars to follow." He did not, however, say when that would happen, only that "I expect to be around to see it."

At the time, though, most people focused on how Obama had changed course, killing Bush's program. But he did throw a bone to the traditional industrial base, pledging to keep Lockheed Martin's Orion crew capsule. Ultimately, after intense negotiations with Congress, the White House would agree to develop a heavy-lift rocket that was similar to the Ares V. The Ares I rocket, however, was gone.

Arguably, the most significant element of the speech, however, was a commitment to rely on the relatively unproven commercial sector for rides to low Earth orbit. This was a fundamental shift for NASA, a move that some in the agency's highest reaches were wary of, and a huge and risky bet for the Obama administration. Even NASA administrator Charlie Bolden was initially against the commercial crew program—putting him at odds with Garver, his deputy. But the White House had made up its mind. Under Obama, NASA would go ahead and retire the space shuttle, and hire contractors to fly missions to the International Space Station. That, in turn, would allow NASA to focus on other missions in deep space.

"Now, I recognize that some have said it is unfeasible or unwise to work with the private sector in this way," the president said during his speech. "I disagree. The truth is, NASA has always relied on private industry to help design and build the vehicles that carry astronauts to space, from the Mercury capsule that carried John Glenn into orbit nearly fifty years ago, to the Space Shuttle Discovery currently orbiting overhead. By buying the services of space transportation—rather than the vehicles themselves—we can continue to ensure rigorous safety standards are met. But we will also accelerate the pace of innovations as companies—from young start-ups to established leaders—compete to design and build and launch new means of carrying people and materials out of our atmosphere."

THE LINE OF the speech, the one that would resonate for years to come, was Obama's "We've been there before." But there was an image that spoke just as loud—one that came about by accident. In addition to the speech, the White House was looking for a photo op, one that would show the president at Cape Canaveral, alongside a rocket, to demonstrate his commitment to space.

To assuage concerns after the battle over Constellation, the White House decided that Obama would visit the United Launch Alliance, the joint venture of Lockheed Martin and Boeing. The message was clear: while the president might have just canceled one of their major programs, the traditional contractors were still a vital part of the American space program. His presence by their rocket would be an endorsement and a signal to Congress designed to ease its concerns.

Only there was a problem. The Alliance was about to launch a highly classified spaceplane known as the X-37B, which would ultimately stay in orbit for 674 days. But doing what? The Pentagon wouldn't say. That was a secret. As was the whole program. Which was why the president couldn't just swing by for a photo op in front of a rocket carrying a highly classified payload. The National Security Council wouldn't hear of it.

So, the White House scrambled. Instead, Obama would visit SpaceX, a high-profile event the company gleefully welcomed. After years of their fighting uphill against the entrenched contractors, a presidential visit would represent a public relations triumph over its archrival, even if it was, as Musk said later, "a sheer accident."

Musk and a small team of SpaceX employees, including Mosdell, greeted the president at Pad 40. They showed him around the launch site, and walked him to the Falcon 9 rocket they had erected on the pad for the photo op. Mosdell couldn't quite believe it. He was walking alongside the president of the United States, while Musk was showing off the pad that he had built.

The photos of that day did everything SpaceX had hoped, and more. The images of the young president walking alongside the young entrepreneur was the greatest endorsement SpaceX could have ever received.

Obama didn't say a word publicly. He took no questions at the pad. And during the speech, he didn't so much as utter the name "SpaceX." But here he was, his jacket slung casually over his shoulder, walking in lockstep with Musk, like two pals out for a stroll. The images were powerful, their message evident: this was the future.

It was as if the president had broken out a bottle of champagne and christened the rocket, blessed its mission, and in the process, tapped the kneeling Musk on each shoulder with his sword, knighting him as a member of the realm.

BUT MUSK SENSED that at one point during their fifteen-minute tour, Obama was also studying him.

The White House had bet big on the rocket now towering over them. It was going to fly cargo to the space station. And it was looking increasingly likely that it would also be one of the rockets NASA would choose to take astronauts there as well.

While it looked majestic standing there on the pad, the fact was the Falcon 9, a much more complicated vehicle than the Falcon 1, had never flown. At the moment, it was little more than a showpiece, an unproven prop in a photo op designed, in part, to shift attention away from the criticisms of the men who had walked on the moon. Given the problems SpaceX experienced launching the Falcon 1 for the first time—and the high failure rate of the maiden flights of rockets in general—Musk couldn't be sure that the Falcon 9's first flight wouldn't end in an explosion.

The president couldn't be sure, either. And Musk couldn't help but feel as though Obama were trying to divine the future.

"I think he wanted to get a sense if I was dependable or a little nuts," Musk said.

The truth was probably somewhere in between.

BEFORE THE FIRST launch of the Falcon 9, Musk found himself in an unfamiliar role: trying to play down the significance of the event. After years of hyping his company, saying it could build more reliable rockets far cheaper, that the future of space lay with SpaceX and companies like it, he was now trying to deflect attention and manage expectations.

It would be a "good day," he said, if just the first stage worked and then the rest of the mission went off course.

"I hope people don't put too much emphasis on our success," he told reporters in the days leading up to the launch. "Because it's simply not correct to have the fate of commercial launch depend on what happens in the next few days. But it certainly does add to the pressure. There's more weight on our shoulders because of that. I wish there weren't."

It was too late for that now. With the White House making a risky gamble that companies like SpaceX could be trusted to fly cargo and eventually astronauts to the space station, far more than just the fate of a single company was riding on the flight. The fate of the industry and a significant portion of the White House's space program was, to a large degree, resting squarely on Musk's shoulders, a burden he and his hard-charging company had put there themselves.

"A dramatic launch failure could further undercut an already faltering campaign by the White House to persuade Congress to spend billions to help SpaceX—and perhaps two other rivals to develop commercial replacements for NASA's retiring space shuttle fleet," a reporter for the *Wall Street Journal* wrote.

On June 4, 2010, less than two months after Musk had toured Obama around the pad, the Falcon 9 was standing vertical again—this time ready to launch, not as the backdrop for a photo op.

On launch day, Mosdell was the launch conductor, in charge of orchestrating all the steps that went into the countdown, and monitoring the health of the rocket and preparing for liftoff. He had worked dozens of launches in his career, but this one was particularly nerve-racking since the rocket had never flown before.

"It was cross your fingers, here we go," he recalled. "In all my experience, I never felt prepared enough. I could always use another day to study this or that, and leading into the SpaceX launches it was that times a factor of ten."

Mosdell was in the back of the launch control room; Musk was up front in the engineering support area, with the vice presidents in charge of propulsion and avionics.

On the pad, the Falcon 9 looked and sounded like a living, breathing animal. Leashed to the tower and its support system, the rocket inhaled vast quantities of propellant in massive heaves and exhaled huge gusts of steam as the liquid oxygen boiled off, like the angry snorts of a bull just before it charges.

Mosdell reminded himself to stay calm and focused on the launch's careful choreography, to take comfort in the precise sequence. To have faith in the script. And, perhaps most of all, to breathe. After every key mile-stone, he told himself, take a deep breath. That would help him get through this. In through the nose, out through the mouth, all the way to orbit.

HE CALLED THE poll, before the launch director declared the Falcon 9 was "ready for launch." There was the T-minus 10–9–8 countdown, and the engines fired. The Falcon 9 lifted off and cleared the tower.

Mosdell took a deep breath.

The engines were humming, shooting out a fiery tail. A little over a minute later, the rocket passed through maximum dynamic pressure, when it was under the most strain.

Another deep breath.

After about two minutes, the first stage engines shut down.

Inhale. Pause. Exhale.

Stage separation.

Deep breath.

Second stage engine started.

Another.

The fairing opened.

One more, as the tension in Mosdell's chest and shoulders began to slowly recede, like the tide going out.

Now he could finally relax. Musk did, too, allowing himself to revel in the latest triumph, the most unlikely of all. While the Falcon 1 had shown SpaceX could get to orbit, it was essentially a test vehicle. And now that the far more advanced Falcon 9 had flown successfully—on its first attempt, no less—Musk declared victory, saying that the launch was "to a significant degree a vindication of what the president has proposed."

It was also a vindication for Musk and SpaceX, one that justified the curious design of the Dragon spacecraft, the spacecraft that ultimately would ferry supplies to the International Space Station. Musk had insisted that Dragon be built with a feature that was completely unnecessary for the passenger-less cargo flights to the station.

The Dragon had windows.

WHILE THE HARE was racing ahead, the tortoise was content to stay hidden in its shell, working quietly deep in the West Texas desert where its secrets were protected. But then on August 24, 2011, a thundering explosion reverberated across the plain, another sign that the supersecretive Blue Origin was up to something.

A little digging would have found that the Federal Aviation Administration issued Experimental Permit Number 11-006 on April 29, 2011. It allowed "Blue Origin to conduct reusable suborbital rocket launches of a Propulsion Module 2 (PM2) launch vehicle" within a 7-mile radius centered on the company's facility in West Texas.

In the days leading up to the launch, it also issued a "notice to airmen" so that airplanes would steer clear of the area.

But the company didn't talk about the launch, and it refused to acknowledge the explosion, officially staying mum, frustrating those looking for answers. Yes, the company's facility was huge and set far, far away from any form of civilization. Not that there was any great population center nearby anyway. But still, that explosion got people's attention. Word spread on social media, and some even called NASA, asking about what had felt like the sky falling. Eventually a reporter from the *Wall Street Journal* caught wind of the explosion and published a story saying the company's rocket had blown up, a setback "highlighting the dramatic risks of private space ventures."

For years, Blue Origin's obsession with secrecy boarded on the absurd. The company was so consumed with staying covert that visitors had to sign nondisclosure agreements (NDAs). Once, even a consultant who wanted to bring his spouse to the company's holiday party was told that, yes, she could come—as long as she signed an NDA, too. Santa himself could have shown up at the party, and the world would have remained none the wiser.

But the rocket explosion had been witnessed—and felt—by more than a few concerned citizens in West Texas, who were beginning to wonder just what the hell was going on in the far reaches of that furtive compound, off Highway 54, where the entrance was marked only by a pair of streetlights and a collection of cameras. For all they knew, they had another Branch Davidian on their hands.

NASA was frustrated. It was now working with Blue as part of its commercial crew program, eventually awarding it contracts worth $25.7 million. After years of being funded exclusively by Bezos, it had become insular, accountable to no one. But now it was working with the government—and blowing up rockets. It had to say something.

David Weaver, NASA's head of communications, called the company's public relations representatives and urged them to make some sort of a statement about the rocket explosion. Secrecy was only fueling the speculation about the mysterious company. It wasn't going to work here. The more time went on, the more it fueled conspiracy theories.

Eventually, more than a week after the explosion, the company published a blog post from Bezos under the headline "Successful Short Hop, Setback and Next Vehicle." It was the first update the company had posted on its site since its 2007 post about the test of the Goddard vehicle.

Bezos led with the good news: "Three months ago, we successfully flew our second test vehicle in a short hop mission," meaning it flew to a relatively low altitude and then flew back, landing safely on the pad.

He continued: "And then last week we lost the vehicle during a development test at Mach 1.2 and an altitude of 45,000 feet." The news there was the company had broken the sound barrier.

"A flight instability drove an angle of attack that triggered our range safety system to terminate thrust on the vehicle." In other words, the rocket was diverging off course, so the engines automatically cut off and it fell back and crashed.

"Not the outcome any of us wanted," Bezos continued, "but we're signed up for this to be hard, and the Blue Origin team is doing an outstanding job. We're already working on our next development vehicle."

He signed it "Gradatim Ferociter!"

In November, the company posted a video of the short hop launch from May, revealing the rocket for the first time. Called Propulsion Module 2, it looked like a farmer's grain silo, squat and primitive. It lifted off in a cloud of smoke and dust, trailed by a fiery tail. It climbed up just a few hundred feet, then momentarily stopped, hovering over the ground just before returning back down ever so slowly, as if it were a marionette being lowered to the stage by the careful hand of a puppeteer.

For decades, the engine was the most important part of the rocket. But this rocket had something altogether different, something that had not been necessary before.

This rocket had legs.

LATER THAT YEAR, on December 2, 2011, Lori Garver got a rare peek behind the curtain at Blue Origin—a personal tour of the company with Jeff Bezos himself.

As they made their way through the cavernous, 300,000-square-foot facility, it was clear Bezos was at home here. He knew people's names, where they had gone to school, what they were working on. The staff wasn't surprised to see him. Here, one of the richest men in the world, the King of Amazon, was Jeff, just Jeff.

"It was a very different experience than your typical CEO tour," Garver recalled.

As someone who was taking heat on Capitol Hill—and within her own agency—for trusting startups like Blue Origin, Garver wanted some first-hand evidence of how the company was different. How it could disrupt the industry. How it could be cheap and reliable.

SpaceX had demonstrated it—Pad 40 alone was a master class in creativity, not to mention the innovative ways it had built its rockets in-house. What, she wanted to know, was Blue Origin's secret?

The answer, in part, was citric acid.

For a while the company had been using a toxic cleaner for its engine nozzles, which it intended to reuse. But that cleaner was expensive and difficult to handle—it had to be used in a separate, clean room because it was so toxic. Then someone discovered that citric acid worked just as well. So, the company started buying it by the gallon, an easier, less expensive solution that worked better.

"Now I'm the largest purchaser of lemon juice in the country," Bezos told her, letting loose one of his trademark cackles.

After about an hour it was clear that Garver, inquisitive, passionate, and supportive, had earned her way into the circle of trust. And so as they sat in a conference room, Bezos leaned in and said, "I want to tell you about my big rocket."

Beyond the PM 2 test vehicle, and even beyond the suborbital rocket that would take paying tourists just past the edge of space, Blue Origin was already sketching out plans for an orbital rocket, one capable of challenging SpaceX's Falcon 9.

Garver wished the company would come out publicly about it. She couldn't help but think of the headlines it would generate, and the support for Obama's space plans. Here was private industry going off on its own to build a new rocket with an American-made engine here in the United States. And it wanted to partner with NASA. That's why Bezos was telling her about it.

But there was no way he would talk about it publicly. It was too early. And part of the company's credo was to only talk about things after they'd been accomplished.

Bezos did, however, invite her to see the company's launch and testing facility in West Texas. That was where it had been testing its engines for the new suborbital rocket, and where the magic happened.

But just because Blue Origin was opening up for the number two at NASA didn't mean it was in any way changing its obsessively secretive culture. The NASA photographer that accompanied Garver on the trip was not allowed on the factory floor. He was forced to wait outside until the tour was over and it was time for the photo op.

He snapped plenty of pictures of Bezos and Garver, along with company executives, including Blue Origin president Rob Meyerson, and Bretton Alexander, its director of business development and strategy. But before the images could be released to the public, company officials insisted they review them.

In the end, they approved just one.

10

"Unicorns Dancing in the Flame Duct"

T HE BIRTH OF the Space Age at Cape Canaveral in the late 1950s was so mesmerizing that the people in nearby, sleepy Titusville were not only starting to believe the United States might just put a man on the moon, but sensing a branding opportunity. "Miracle City," they'd call their town.

Following the government's infusion of Cold War cash, the community had grown from a population of 2,604 in 1950 to more than 30,000 by 1970, and the developers of Titusville's new 330,000-square-foot Miracle City Mall were ready. "Miraculous profits await you," boasted a brochure designed to attract new tenants to a shopping center described as "as modern as the space age activities of its neighbors."

"Miracle City" may have sounded like the slick marketing department hype of a coastal Florida real estate development firm. But it was apt. What was happening along this stretch of quiet beach was indeed miraculous. NASA didn't even exist until 1958. Three years later, after the United States had built a space program almost from scratch, Alan Shepard had become the first American to reach space. A decade later, he would hit a makeshift 6-iron, smuggled onto Apollo 14, in a lunar dust trap.

With unprecedented investment throughout the 1960s, NASA put on an amazing show, building new rockets and spacecraft, training a generation of astronauts who would pull off the impossible, and, while beating the Russians to the moon, inspiring the world. For the backdrop to this improvisational drama, NASA had built a suitably grand stage: Launch Pad 39A.

It stood like a skyscraper on the Florida coast, its spire stretching nearly 500 feet high. Before launch, the astronauts zipped to the top in an elevator, getting one last view of the waves lapping the earthly coastline. And there, atop the scaffolding, just before the bridge that took them to the rocket, there would be a telephone for the astronauts to make their final calls, as if they were facing a prison sentence. Like a child's toy, the phone had extra-large buttons, all a shiny gold, designed specifically for astronauts outfitted in bulky spacesuits and gloves.

If Launch Pad 39A was the stage, the star of these explosive performances was the Saturn V, a monster of a rocket with five engines—hence the V—that generated enough force to power New York City for more than an hour, consuming fuel at a rate of 15 tons per second. Fully fueled, the Saturn V weighed more than 6.2 million pounds. It had 3 million parts and to this day remains the most powerful rocket ever built. At ignition, flames and thick billowing plumes of smoke gushed from its engines, each nearly two stories tall, and surged through a flame trench the size of a subway tunnel. The roar reverberated like an earthquake for miles, and the people of Titusville joked that they weren't sure whether the Saturn V took off or that Florida had suddenly sunk into the ocean.

This was where NASA would showcase many of its most important launches. It was the site from which Neil Armstrong, Buzz Aldrin, and Michael Collins blasted off toward the moon in 1969. Then in 1972, the pad was where Eugene Cernan, the last man to walk on the moon, took off. Liftoff after liftoff, 39A became the Broadway of the Space Age, an amphitheater large enough to fit even the most outsize ambitions. In 1981, it launched the first space shuttle into orbit. Thirty years later, it would host the very last shuttle departure, marking the final chapter of an extraordinary era of human spaceflight.

But in 2011, the retirement of the shuttle came with a jolt, and was met with widespread disbelief—and denial—on the Space Coast, which could not fathom the new, hard truth: After fifty years of historic launches, the United States was suddenly, for the first time in decades, incapable of launching astronauts into space. Instead, the nation would have to rely on Russia—which it had bested in the race to the moon—for rides to the cosmos.

The dreams of Apollo were decaying, as were its stomping grounds.

Left behind on the Cape were the ruins from a once-great human space program—abandoned spires of launch towers, grown over bunkers that once housed launch crews—clues for future archaeologists of what had once transpired on this sacrosanct ground. Then there were the artifacts that were no longer visible, the rusted launch site skeletons that had been torn down or buried, their existence only hinted at by the paths beneath the brush leading to nowhere.

At Launch Pad 14, not far from 39A, a forbidding gate with a curious sign proclaimed it protected the "launch site of the free world's first man in orbit." That would be John Glenn, who had mimicked the feat performed by Yuri Gagarin, the first man in space, the year before for the Soviet Union.

The tour buses stopped coming, and so rarely anyone saw the museum-like exhibit, faded by the relentless Florida sun and clouded by the salt air. It explained all that had happened here at Pad 14, which began not with John Glenn, but with a chimpanzee named Enos, from Cameroon, who was trained as an "astrochimp" at the University of Kentucky and Holloman Air Force Base in New Mexico.

"He became the first living creature launched by the U.S. to orbit the Earth when he flew aboard Mercury-Atlas 5 on November 29, 1961," the exhibit read. "Enos logged a total of 3 hours, 21 minutes in space and paved the way for the first U.S. manned orbital flight just three months later."

Inside, just before the overgrown launchpad, there was another curious clue in the parking lot. Four parking spots each had a plate bearing the name and military rank of one of the four Mercury astronauts who had launched from Pad 14—"John H. Glenn Jr. LT. COL.," "M. Scott Carpenter LCDR," "Martin M. Schirra Jr. LCDR," and "L. Gordon Cooper MAJ."

The parking spots, though, sat empty and waiting, as if the ghosts of the astronauts would one day return.

LAUNCH PAD 39A, left dormant, was rusting away in the salt air. Weeds grew in the flame trench, spurts of new green grass breaking through the remnants of char. Up in the scaffolding, the astronauts' phone sat forgotten, with no one on the other end to receive its calls. Its oversize gold buttons faded to brown.

Outside the gates of Kennedy Space Center, the Space Coast faltered, the underpinning of its economy gone. The Miracle City Mall inevitably succumbed, too. By the last shuttle flight, it was home to just two businesses, JC Penney and a hot dog stand. Eventually, it was torn down.

After standing as a monument to American ingenuity and the embodiment of John F. Kennedy's lunar aspirations, 39A was now a symbol of the degradation of the US human spaceflight program. To keep down maintenance costs, which were $100,000 a month and climbing, NASA had already disassembled much of it. And a spokesman admitted, "It's not in a hazardous situation, but it has not been kept up."

Now, in 2013, some forty years after it was erected, NASA wasn't sure what to do with the launchpad. The structure was on the National Register of Historic Places, and couldn't be torn down.

Unused, 39A was now a tower of obsolescence, a burden to NASA and taxpayers and a painful reminder of past glory. The only solution was for NASA to find someone to take it over. Someone crazy enough to be willing to spend the money to strip it down to the studs and breathe new life into a faded beauty.

NASA officials knew not too many people were in the market to lease a fixer-upper launchpad, but they had their eye on one possible tenant: an eccentric billionaire who had started a space company from scratch with absolutely no experience with rockets, but talked about colonizing Mars—a wild card named Elon Musk, who was now on an improbable, but epic, roll.

THE FALCON 9 had flown successfully. And SpaceX was moving ahead with developing a more robust version of its Dragon spacecraft that would carry astronauts, not just cargo. It was talking about building an even bigger rocket, called the Falcon Heavy, which would allow it to pursue Musk's original goal of colonizing Mars. Musk even put a price tag on it, telling the BBC, "Land on Mars, a round-trip ticket—half a million dollars. It can be done."

In May 2012, the company was aiming for yet another major milestone when it flew its Dragon spacecraft to the International Space Station.

Launching a rocket was one thing; flying a spacecraft to orbit and having it berth, or dock, with the orbiting station was a far more difficult feat, one that had been accomplished only by three countries—the United States, Russia, and Japan.

The pressure to get it right was enormous, and some SpaceX employees had been working nonstop for months. In the hours leading up to the Dragon's arrival at the station, one exhausted engineer, who had been up all night, slumped against a wall in the company's headquarters holding a sign, like a homeless person begging for change: "Hungry and tired. Please berth."

Flying above Australia, American astronaut Don Pettit maneuvered the International Space Station's 57-foot-long robotic arm to reach out and grab Dragon, the world's newest spacecraft. As the orbiting laboratory careened around the globe at 17,500 mph, the astronauts aboard the station carefully guided the Dragon capsule into position, making it the first private company to accomplish the task.

"Looks like we've got us a Dragon by the tail," Pettit told the NASA administrators in Houston.

At SpaceX's headquarters just outside of Los Angeles, employees broke into raucous applause, chanting their boss's name, "We love Elon!" Musk was now developing a cultlike following, and SpaceX had swelled to more than two thousand employees with an average age of thirty, with $4 billion in contracts.

"This is, I think, going to be recognized as a significantly historical step forward in space travel," he said afterward. "Hopefully, the first of many to come."

ON MARCH 1, 2013, the Falcon 9 lifted off for its second official cargo delivery to the station. While the rocket did so smoothly, within an hour it was clear that its Dragon spacecraft was in trouble.

"It appears that, although it achieved Earth orbit, Dragon is experiencing some type of problem right now," John Insprucker, SpaceX's Falcon 9 principal integration engineer, said on the company's webcast before signing off. "We'll have to learn about the nature of what happened."

Inside mission control, the SpaceX team was desperately trying to figure out what was wrong, and soon pinpointed the problem: a valve was stuck.

Steve Davis, SpaceX's director of advanced projects, started to prepare for the worst—aborting the mission entirely and bringing the spacecraft back to Earth. But the crews wondered, "Is the vehicle even functioning enough that you can bring it back?" he remembered. "We weren't sure. That was the only time we had ever planned for an emergency re-entry, which is like a big thing because you have to whip it through air space. You have to reroute planes in real time. It's not awesome. And so we were in that panic mode."

They had been in that panic mode before. In late 2010, on the eve of the Falcon 9's second launch, and the first test flight of the Dragon spacecraft, a last-minute inspection of the rocket revealed a crack in the nozzle, or skirt, of the second-stage engine. That was not good.

"You're not going to fly with a crack," Davis said. "We're like, 'What do we do?'"

The normal thing would be to take the rocket apart, replace the engine skirt, reinspect it, and then "you're up and launching in a month," he said. No one wanted to lose that much time.

Instead, Musk had a wild idea that he put to his team: "What if we just cut the skirt? Like, literally cut around it?" That is, what if they trimmed off the bottom as if it were a fingernail?

"He went person by person and said, 'Would this have any adverse effect on you?'"

Davis said that because the skirt would be shorter, they'd get less performance from the engine. "But we had so much margin built into it, it didn't matter." Everyone else concurred, and "literally within thirty minutes, the decision was made."

The company flew a technician from California to Cape Canaveral; armed with a pair of shears, like the kind used to trim hedges, he cut around the crack. "And we flew the next day successfully," Davis said. "That could have been the dumbest thing we ever did, but it was amazing."

That was not how NASA would have handled it. But its officials agreed with SpaceX that there wasn't any reason why it wouldn't work, and approved the launch, astounded by how quickly SpaceX was attacking the problem.

Now, as Dragon was in trouble with the stuck valve, NASA was similarly hands-off.

Bill Gerstenmaier, NASA's associate administrator for human exploration and operations, and Michael Suffredini, the space station's program
office manager, were in the launch control center watching over SpaceX's
shoulder as they tried to figure out how to "burp the valve." They were
two of the agency's most senior officials, with nearly sixty years at NASA
between them. They had served through the shuttle disasters, had seen
all sorts of problems in space, and now, as NASA faced another potential
crisis, they were just talking softly between themselves.

Standing nearby, Lori Garver, NASA's deputy administrator, could
barely contain herself. SpaceX's Dragon was in trouble—deep trouble, it
seemed. If it didn't dock with the station, if the mission somehow failed,
the critics would come out again to blast Obama's decision to rely on
these contractors. This mission had to work. They had to find a way to
rescue Dragon, and fast.

There were no better people to come fix this than Gerst, as he was
known, and Suffredini. But there they stood, two of NASA's elder states-
men just watching, offering a bit of advice, a whisper here, a suggestion
there, to the SpaceX kids—and really, they looked like kids. But mostly,
both were staying out of the way, letting the kids figure it out.

Garver desperately wanted them to take over, to swoop in and save
SpaceX. But instead they stood back.

"They were like grandparents instead of parents," Garver recalled.
"And it was almost like grandpa taking them fishing: 'Try over there.
There might be some fish over there.'" A soft touch designed to let the
kids learn to fish on their own, rather than an impatient dad's just grabbing the pole and catching the fish for them.

"If there was something we saw that we could have interjected, we
would have done it," Gerstenmaier recalled. But it wasn't NASA's spacecraft. The wise elders weren't in control.

"We really were in an advisory role," Suffredini said. "We couldn't give
them any help but high-level guidance."

As they watched, the kids in the control center were making progress
on the problem. The valve was stuck, so they'd need something to make
it unstuck. On a spacecraft circling the globe at 17,500 mph, that was no
easy task. But the SpaceX team knew that if pressure could be built ahead
of the valve, and then it was suddenly released, that might just deliver the
kick needed to jar the valve open.

"It's like the spacecraft equivalent of the Heimlich maneuver," Musk said later.

One of the engineers wrote a command, right then, on the fly, programming the spacecraft to build up the pressure. Then, they tried to beam the new command up to the Dragon, as if it were an iPhone update. At that moment, the folks at NASA knew they were witnessing something special. It wasn't that they had fixed a problem with the spacecraft; that happened all the time. It was how fast they did it.

"The SpaceX mind-set had always been about adapting quickly, and it really shined that day," Suffredini said. "They had really an in-depth understanding of that system and the software, and that's one of the secrets of their success. They probably had the kid in there who wrote the original code."

But the SpaceX crew was having a hard time communicating with the spacecraft. The code wouldn't transmit. So, someone got the air force on the phone, which then gave the company access to a more powerful satellite dish, which allowed, at last, the uplink.

The code worked. The valve opened. Dragon was able to dock with the station.

The whole SpaceX team could take a deep breath.

"That was nerve-racking," Musk said later. "For a while there, we thought we'd have to abort the mission. But we were able to upload new software and make it work."

Watching from a distance, the grandparents were pleased. SpaceX had caught a big fish all on its own.

It was a turning point, SpaceX's rite of passage, an entry into adulthood. It had matriculated and was now granted membership into the rarefied, old-boys club. But nothing would symbolize the apotheosis of SpaceX, and by decree the emergence of the New Space movement it was now clearly leading, except the transfer of 39A into private hands. By 2013, Musk had set his sights on winning 39A, and it seemed inevitable that he would add the world's most venerable launchpad to his growing trophy case.

As far as some in NASA's leadership were concerned, the agency should just sign the pad over to SpaceX, no questions asked, and be grateful that someone was willing to take it over. But others knew that would be a problem. They needed to bid it out, if only for the sake of procedure.

Of course, they'd get no other takers. Who would want a used launchpad that would require millions to restore?

Then, from out of the blue, NASA received another bid, this one a surprise from a little-known company that had for years remained quiet and secretive. But now Blue Origin was slowly emerging from the shadows.

A YEAR EARLIER, in October, Garver had visited Blue Origin again, this time at its facility in West Texas. By now, almost a decade after Bezos had begun buying up the land in secret, there was a fully functioning launch facility, complete with an engine test stand and a launch site.

Garver was particularly interested in the test stand, since NASA was looking into refurbishing one of its own at a cost that seemed astounding to her: $300 million. As she toured Blue Origin's test stand, she asked her guide, a young engineer, whether he knew how much it had cost to build. His answer: somewhere in the range of $30 million.

Ten times cheaper. She was stunned, and reminded again of how efficient private industry could be. "Can we test our rockets here?" she asked the people at Blue, who demurred, not wanting to deal with the government bureaucracy.

In addition to the test stand, Blue had an odd chart to show her. On the wall in one of the offices was a grid laid out with a series of squares. Each square represented a plot of land in the Texas desert. In a couple of days, the company was going to perform what was essentially a fire drill in rocketry—a pad escape test. The idea was to demonstrate that if anything went wrong with the rocket, the capsule sitting on top, holding the astronauts, would be able to fly away to safety.

For the test, they'd put the capsule on the launchpad, and then fire its motor to ensure that, in the case of a problem with the booster, the spacecraft would be able to get the crew away as fast as possible. The grid that Garver was now looking at represented the spots where Blue's employees thought it would land, after flying a couple thousand feet into the air and then parachuting back safely to the ground. The Blue Origin team made a game of it. For $5 you could pick a square, and then if the capsule landed there, you'd win.

Garver made her pick, but told them that if she won, she'd donate the winnings back to the team. A few days later, on October 19, 2012, Bretton Alexander, Blue's director of business development, sent her an e-mail.

"Success!!" read the subject line.

"Awesome pad escape test!" he wrote. "Need to review data but looked beautiful!"

"Oh my . . . congrats! I was just going to write u!!!" she responded.

"Btw you are one of 11 winners who picked the winning square!! We were 11 feet from the predicted mean location!! And we've spent your winnings on beer, scotch and tequila ;)" Alexander wrote back.

"You got a great shout out for your game AND accuracy :)."

A FEW MONTHS later, in January 2013, Blue had some more news. In an e-mail, Rob Meyerson, the company president, wrote to NASA administrator Charlie Bolden and Garver to say that the engine that would power its New Shepard rocket was making significant progress.

"My apologies for not making it up to the Museum of Flight today for your visit," Meyerson wrote on January 15. "I had planned to go (even had my coat on at one point) but decided to stay back in Kent so I could watch the first test of our new BE-3 engine, which was conducted at our West Texas site. The BE-3 is a Blue-developed 100,000 lb thrust rocket engine that uses liquid oxygen and liquid hydrogen as its propellants. After overcoming the usual first test obstacles, we were able to get the test off around 4:00 pm today. It was a tremendous milestone for Blue Origin and the result of many years of effort."

He went on to thank NASA for its support, saying it helped the company save about a year in development time.

This was indeed a tremendous milestone—and huge news, a Henry Ford moment: Jeff Bezos was building a rocket engine. Garver immediately sensed a public-relations opportunity for NASA and the White House. Since they had backed Blue with $25.7 million in contracts, and were supporting the private space industry, she wanted to shout this success to the rooftops. Let all those doubters in Congress, in industry, even in NASA's own leadership, know that these companies, with help from the government, could succeed.

"Your note about NASA's assistance saving you a year of development time is especially welcome," Garver wrote to Meyerson. "I'd really like to be able to communicate that message more broadly in upcoming speeches, testimony etc. Are you folks open to coordinating on such a message?

"I know you are the 'quiet company,'" she continued, "so I don't want to presume to be able to share the information. And either way, it is really wonderful to see the government/industry teams working with such synergy."

There would be an announcement, but not until more than a month later. And the news of the engine test was mentioned only in passing, as part of a broader press release from Blue about how it was continuing to test its engine through a partnership with NASA, though without receiving any additional funding.

Through 2014, the company planned to continue to test its rocket and capsule methodically, "putting emphasis on power and actuation systems, in-space propulsion, multiplex avionics and flight mechanics. The company also will progress the spacecraft's guidance, navigation and control systems."

In other words, Bezos was getting ready to fly.

BEZOS WANTED 39A for the new rocket it was developing in secret that went by the nickname "Very Big Brother" in-house. The launch site was a national treasure, one he had been fascinated with since he was a five-year-old kid watching the crew of Apollo 11 take off, a "seminal moment for me," as he later commented. If Musk won the exclusive rights to it, it would be as if NASA was saying it had chosen SpaceX as the rightful heir of Apollo.

Blue Origin had remained content to be on the sidelines for the better part of a decade. But no longer. That silence ended now. Launch Pad 39A, and all that it represented, were too big a prize. If NASA was giving it up in its will, Bezos would make a bid.

Bezos's team had tried to win the rights to 39A by arguing in 2013 that the venerable pad should not be operated exclusively by any one company. Unlike SpaceX, Blue Origin promised to share it with others, such as Boeing and Lockheed Martin and even SpaceX.

NASA looked at both proposals, and studied the pros and cons. Musk and NASA had already had a long-running relationship. NASA was investing billions in SpaceX. Even President Obama had blessed the company, if implicitly, by visiting Pad 40, a few years before.

The fact was, Blue Origin didn't yet have a rocket capable of launching from 39A. The hare had sprinted far, far ahead. The tortoise's slow, deliberate approach might, one day, allow it to catch up. But now, it was too far behind. The competition wasn't even close. Musk won, hands down, adding the iconic pad to a long list of triumphs, which now included besting Bezos in their first high-stakes head-to-head clash.

It might have ended at that. But Bezos wasn't about to give up. Blue Origin sought to reverse the award by filing a legal protest, arguing that the criteria NASA used to come to its decision were flawed. It argued that the launchpad should be a "commercial spaceport" that several companies could use.

Then to bolster its case, Blue Origin enlisted the support of the United Launch Alliance, the joint venture of Lockheed Martin and Boeing—SpaceX's chief competitor, which was eager to join the fray in a move it knew would only antagonize Musk.

The Alliance jumped wholly into a nice, convenient partnership with Blue, a marriage that brought together the heritage of a legacy contractor, with the innovation of a new startup—not to mention one backed by one of the wealthiest men in the world. In a statement to *SpaceNews*, the Alliance said that it would "continue to share our technical expertise in launch infrastructure with Blue Origin," which, in turn, would allow 39A to have multiple tenants.

It enlisted the aid of friendly senators, who in a letter to Bolden, the NASA administrator, wrote that "blocking use of the pad to all but one company would essentially give that company a monopoly, stifling competition in space launches and therefore raising costs."

The legal protest was a case of "launch site envy. That was annoying," Musk said later. "Filing a lawsuit for 39A when they haven't even gotten not so much as a toothpick to orbit. . . . So, it was absurd for [Bezos] to claim that Blue Origin should get 39A."

The protest, the lobbying in Washington, and the sudden, back-channel union between the Alliance and Blue infuriated Musk, who had also grown agitated that Blue Origin was beginning to poach some of his employees. In his biography of Musk, Ashlee Vance reported that SpaceX even designed an e-mail filter designed to search employees' e-mails for "Blue Origin."

The dispute over 39A wasn't their first entanglement. In 2008, SpaceX sued Matthew Lehman, one of its former employees, alleging he had violated his contract—that Blue Origin used the information he provided "to attempt to recruit multiple SpaceX employees with specific and detailed knowledge of SpaceX's design efforts and of extensive confidential SpaceX information relating to those design efforts," the lawsuit claimed. "Blue Origin utilized extreme measures to entice these carefully targeted SpaceX employees to leave their SpaceX employment and join Blue Origin."

The lawsuit was eventually dismissed. But the tensions lingered. And now the fight over 39A only inflamed them. Musk fired off an e-mail to *SpaceNews* in September with an epic takedown of his new competitor, deriding the protest as a "phony blocking tactic and an obvious one at that." Even though Blue Origin had been around for a decade, it "has not yet succeeded in creating a reliable suborbital spacecraft," he wrote.

"It is therefore unlikely that they will succeed in developing an orbital vehicle that will meet NASA's exacting standards in the next 5 years, which is the length of the lease. That said, I can't say for sure whether [Blue Origin's] action stems from malice. No such doubt exists about ULA's motivation."

There was a subtle jibe laced in there, one that most people would miss but that was of great importance to Musk. As he would point out again and again over the years, Blue's New Shepard rocket would be suborbital, and therefore not nearly as powerful as the boosters he was building that were capable of reaching orbital escape velocity, the speed needed to break Earth's gravity and stay aloft, in orbit. Blue's New Shepard, by contrast, would go up, and then come straight down, like a ball tossed into the air.

"However, rather than fight this issue, there is an easy way to determine the truth, which is simply to call their bluff," Musk continued. "If they do somehow show up in the next 5 years with a vehicle qualified to NASA's human rating standards that can dock with the Space Station, which is what Pad 39A is meant to do, we will gladly accommodate their needs. Frankly, I think we are more likely to discover unicorns dancing in the flame duct."

Unicorns in the flame duct. Whether he meant it to be or not, this was a rallying cry to his troops, who delighted in how bold their leader could be. There was, however, an irony: Musk was treating Blue the same way Boeing and Lockheed had treated SpaceX a decade earlier, when it was filing lawsuit after lawsuit, trying to enter the market. The legacy contractors had derided SpaceX, calling it an "ankle biter," saying it wasn't a serious challenger because it did not have a proven rocket.

Blue could have responded to Musk's taunt by announcing that it was also developing an orbital vehicle, "Very Big Brother," thank you very much, one that would be powered by new engines, made in-house. But it didn't take the bait. Musk's insult didn't produce anything but more of the same obsessive, disciplined silence.

Bezos was sticking to his own advice written in Blue's founding letter a decade earlier: "Be the tortoise and not the hare."

VICTORIOUS, SPACEX STARTED work immediately, renovating 39A. The Kennedy Space Center was to space what the White House was to politics. Only now, one of the jewels of the space center had a corporate insignia on it, a giant SpaceX logo spread across the side of a massive warehouse. It hadn't yet gotten to Mars, but SpaceX had planted its flag on some of the most sacrosanct soil of the Florida Space Coast.

The company was moving on, looking to the future. But the spat with Blue Origin left many at SpaceX angry. The launch site director took a photo of about one hundred blow-up unicorns that he had placed in the flame duct. And in a tucked-away conference room in its Washington office, there was a picture of Captain Jean-Luc Picard, the star of *Star Trek: The Next Generation*, a follow-on to Bezos's favorite childhood television show.

A bubble coming out of his mouth read: "What the fuck does Blue Origin need a Florida launchpad for?"

PART III

INEVITABLE

11

Magic Sculpture Garden

J EFF BEZOS BLAMED the bananas.

In early March 2013, he had quietly stolen away from his growing Amazon empire for a three-week expedition at sea, with a team of some of the best deep underwater ocean explorers in the world. Yet despite its vast experience, the crew had somehow violated one of the oldest seamen's superstitions: never bring bananas on a boat.

The crew of the *Seabed Worker*, a Norwegian salvage ship outfitted with the most advanced underwater robots money could buy, had packed in bananas by the bushel. And now, as the ship was getting battered by a late winter storm the Weather Channel had dubbed "Saturn," the curse of the bananas had come to haunt them.

Saturn had swept from the Rockies east through the Midwest, coating a vast portion of the country in snow before coming out at sea, where its winds were churning the Atlantic Ocean into frothy, menacing waves. Some 15 miles off the coast of Maryland's Assateague Island, the storm pummeled a 67-foot fishing vessel, shredding the hull and killing two members of its three-man crew.

Soon it was rocking the *Seabed Worker* and its crew of sixty, which included not only the all-star team of explorers Bezos had assembled for this secretive mission, but several members of Bezos's family as well: his parents, brother, and brother-in-law.

This was supposed to be a family-friendly trip of adventure and discovery, a quest for lost treasure on the high seas. But then the storm stalled overhead. The winds howled and waves towered over the ship, spraying in thick, drenching sheets that cascaded over the hull as the ship bobbed to and fro like a metronome.

They considered trying to escape to the south. But the storm was too big. They had no choice but to ride it out.

IT WAS A slow day in July 2010 when David Concannon got the call in his law office in Philadelphia's Main Line. The caller would only give her first name, and though she said she was calling on behalf of a client, she didn't say who it was. Concannon had been getting a lot of crazy calls lately—just the other day, someone had urged him to go search for a secret fort, down by the airport, which had been built by the Knights of Templar, a twelfth-century Christian military order.

This woman seemed like just another crackpot. "But I was bored that day, so I talked to the person," he recalled. Finally, she allowed that she worked for a "high net-worth individual," and came out with her question: Would it be possible to recover the F-1 engines from the bottom of the Atlantic Ocean?

Concannon had no idea what she was talking about. He Googled it and realized that F-1s were either from a racecar or the Apollo-era Saturn V rockets. The former meant nothing to him; but recovering the engines from the rockets that took astronauts to the moon, well, that would be a coup, especially for an explorer like him.

In addition to being an attorney, Concannon ran a consulting firm for explorers. For years, he had helped put together expeditions to Mount Everest, and to the bottom of the ocean, where his teams had helped recover artifacts from several sunken ships, including the *Titanic*.

"Yeah, it's possible," he responded. "Anything's possible."

BUT IT WOULD be exceedingly difficult, as he discovered researching what it would take—even more challenging than finding the wreckage of the *Titanic*. The *Titanic* was 883 feet long and weighed more than 52,000 tons. On the ocean floor, it stood six stories high. The explorers that found it in 1985 had decent information on where precisely it sank. And, perhaps more important, previous expeditions had searched hundreds of square miles, so they knew where not to look.

The F-1 engines, by contrast, were tiny. "Like finding a deck chair or a boiler on the *Titanic*," Concannon later recalled.

No one knew where they were, at least not exactly. Once the Saturn V took off from the Kennedy Space Center's Pad 39A, its first stages separated and fell back to Earth, where they eventually splashed down into the ocean. But no one had tracked their descent. NASA had only a vague sense of the impact point based on the flight path. But the space agency didn't track them on radar, and didn't even issue a "Notice to Mariners" warning ships to steer clear of the area where a rocket booster would be falling from the sky.

Even if the engines could be found, it was unclear what kind of condition they'd be in. Maybe, red hot, they'd torn apart when hitting the cool Atlantic, shattering to pieces. Maybe after forty years on the bottom of the ocean, they had deteriorated away. When Concannon first saw the wreckage of the *Titanic* sitting off the coast of Newfoundland at a depth of 12,460 feet, where the pressure is more than 6,000 pounds per square inch, he was shocked not by its grandeur but rather by its "appalling condition."

"The *Titanic* looks like it is made of wet sand," he once wrote in an account of the mission. "The ship looks nothing like I imagined. Instead, it appears to be rotting away, like a candle melting from the top down. It's easy to believe that the *Titanic* will be nothing more than a stain on the ocean floor in a few years."

The Apollo engines could be in similar condition, or worse. And that was assuming they could be found.

Whoever was behind this mysterious phone call would have to have immense amounts of patience, not to mention money. The person would also have to be comfortable with attempting a mission most people would classify as foolhardy, if not impossible.

A MONTH LATER, the mystery woman called back. Her boss was interested in proceeding, and now she was ready to reveal his identity: Jeff Bezos. Concannon wasn't surprised. He was not aware that Bezos had interest in space or even ran a space company, but he had worked with lots of rich people and even some celebrities, including James Cameron's *Titanic* trip, to know that wealth often fueled eccentricity.

To Bezos, the engines were the embodiment of the Apollo missions that had inspired him as a five-year-old. He was fascinated with the F-1s, their brute power, and called them a "modern wonder." The engines had 1.5 million pounds of thrust each; all five together burned through 15 tons of fuel per second, and fired for just two and a half minutes before the rocket's first stage separated and fell into the ocean.

"It's hard to find something more profound as an engineered object than the F-1 engine, the most powerful single chamber rocket engine ever designed and manufactured," he said. "Sixty-five of these engines flew, and there were zero failures."

While they might have been hunks of steel to others, with no real significance, to Bezos they were important artifacts. "These are the actual objects that first took humankind to the moon," he said. "They're a marvel—for me, they conjure the memory of the thousands of passionate engineers who brought the Apollo program into existence."

But they had been sitting on the bottom of the ocean for more than four decades, "and they weren't going to last forever down there," he said.

At the time, Blue Origin's New Shepard rocket was well under development. Compared to the 363-foot-tall Saturn V, it was a pipsqueak—just 65 feet tall. Its single BE-3 engine was capable of just 110,000 pounds of thrust, compared to the F-1's 1.5 million.

But the little New Shepard was being designed to do something that the Saturn V could not. It would be able to launch to space, and then fly itself back to Earth, autonomously, adjusting its course on the way back down, able to guide itself precisely back to a landing pad, so that it could fly all over again and again.

To Bezos, then, the F-1s not only represented the Apollo era's gargantuan feats of engineering—but would also be reminders of how rudimentary they were, of a time when rockets were expendable, never to be used again. Once recovered, they'd be showcased where they belonged: a museum, representing perhaps humanity's most historic achievement. But hopefully, they would also come to be seen as a relic, as antiquated as a horse and buggy.

"Blue Origin is determined to write a new chapter—reusability," Bezos would later say. "That's the key to making space travel affordable. No more throwing engines away in the ocean. We don't want anyone to be able to recover our engines from the Atlantic fifty years from now!"

On September 24, 2011, Concannon's crew took off from Newport News, Virginia, in a former navy spy ship, the 224-foot-long *Ocean Stalwart*, which had been overhauled as a research and surveying vessel.

This was a reconnaissance mission, dispatched by Bezos in an attempt to see whether they could find the engines. The ultimate goal was to find an engine used during the Apollo 11 mission; Bezos wanted a piece of the historic lunar landing. If the crew was successful, it'd come back at a later time in a separate expedition with Bezos to recover what they had found. The 180-square-mile search area, a few hundred miles off the Florida coast, where they thought they would have their best luck, was going to be at extreme depths, more than 14,000 feet, about 2 1/2 miles down—deeper than the wreckage of the *Titanic*. There the ocean floor was like a moonscape, ghostly and largely devoid of life. Sunlight didn't penetrate at that frigid depth, making it completely black. The water pressure was 7,000 pounds per square inch.

Instead of searching the ocean floor with cameras, they'd use sonar, pulses of sound that would reflect off objects on the seafloor and measure their distance and bearing. Sonar would allow them to scan greater distances, as much as 4,000 feet, or even something as small as an airplane propeller from a distance of more than ten football fields.

But getting the sonars to the seafloor was a tremendous undertaking. The ship had to be able to keep its position, without an anchor, by using its engine thrusters while being pushed around by the currents, waves, and winds. The side-scan sonar system affixed to a 15-foot-long, 6-ton "tow fish" that looked like a mini torpedo. The tow fish were lowered to the bottom of the ocean by a spool of cable 32,800 feet long that, by itself, weighed more than 20 tons. In all, the sonar system cost more than $1 million.

Once the boat was in position and ready to search, it took five hours to lower the tow fish into position, where it would begin scanning the ocean floor, identifying anything that could be a rocket engine. Over two weeks, with crews working around the clock, the team created a map of the ocean floor across the search area. The painstaking effort paid off: they found thousands of man-made objects and more than three hundred so-called high-value targets they felt could be rocket engines, clustered in eighteen different areas.

After studying the data, Bezos announced the discovery several months later in a blog post: "We don't know yet what condition these engines might be in—they hit the ocean at high velocity and have been in salt water for more than 40 years. On the other hand, they're made of tough stuff, so we'll see."

The engines were found. Now they had to get them up to the surface.

In February 2013, as the White House was fighting to kill the Constellation program, and Musk was about to meet President Obama at Pad 40, a Norwegian ship took off from Bermuda, ready to recover the engines.

Funded entirely by Bezos, it had the best of everything in what was surely a multimillion-dollar undertaking. Concannon had assembled an all-star team, including John Broadwater, the former chief archaeologist at the National Oceanic and Atmospheric Administration, and Vince Capone, one of the world's leading underwater search experts. The team brought along a doctor as well, Ken Kamler, a specialist in expedition medicine who had served on many perilous adventures, including to Mount Everest, where he treated the survivors of the trip chronicled in Jon Krakauer's *Into Thin Air*.

In all, about sixty people were aboard, including Bezos himself, who would spend as much as three weeks at sea, away from his Amazon empire, though he'd spend stretches in his cabin with his computer working during the downtime.

Later he would say, it was "so fun" having members of his family along. "It was meaningful to share the experience" with his parents, Jackie and Mike; his brother, Mark; and his brother-in-law, Steve Poore. Since Jackie was the only woman on board, the captain jokingly told Bezos that he would remove all pornography from the ship's common areas.

For this mission, the team acquired a recovery ship known as the *Seabed Worker*, a marvel that towered six stories over the ocean, was 290 feet long, and weighed almost 4,000 tons. The cockpit, or bridge of the ship, looked like the command center of a massive spaceship, with a large, plush chair for the captain as well as joysticks and several computer screens that relayed all sorts of data in real time. And it was outfitted with what's known as a dynamic positioning system, which uses GPS to keep the ship precisely over a target.

Perhaps most important, the *Seabed Worker* had two remotely operated vehicles (ROVs), essentially underwater robots that could be controlled from the ship. These $7 million ROVs were able to work at extreme depths.

"We're working three miles down in the deep ocean," Capone said. "And our robots are like marionettes on three-mile-long cables. Being able to control those robots is a fantastic deep water ballet."

The weather was rough as the ship powered its way 500 nautical miles southwest from Bermuda to the recovery site, but the *Seabed Worker* was a gallant trooper, able to absorb the waves, allowing the crew to operate the ROVs in seas as high as 15 feet. By the time they got to the search site on March 2, 2013, the weather had calmed. They were ready to start plucking metal from the depths. "You can feel it walking around the ship everyone's excited," said Jeff's brother, Mark. "There is some level of trepidation because we know there are no guarantees."

The ROVs made it to the seafloor at eleven a.m., Sunday, March 3, and almost instantly beamed back images of engine parts that were broadcast in high definition on large-screen TVs installed just for this mission. "We found the first artifact within minutes and found the first engine within the first hour of that first day," Concannon said.

There on the ocean floor were the thrust chamber, turbo pumps, and heat exchangers. To some, they'd look like twisted hunks of metal, fit for a junkyard. To Bezos, they were art and history combined.

"Three miles below where I'm standing right now is a wonderland that is testament to the Apollo program," he said in a video recorded on board the ship. "It looks like a magic sculpture garden, with all of these pieces from different missions that are in some cases perfectly preserved and in other cases twisted into the beautiful shapes."

Within a couple of hours, they found even more about 330 feet away. One was buried into the seabed floor so deep they nicknamed it "lawn dart." Over the next couple of days, they surveyed the area, taking photographs, cataloging location and conditions. Meanwhile the weather was deteriorating and the seas were rising. Winter Storm Saturn was moving to the east, and they knew they had to move fast.

"It was very rough when we got there. And it had been very rough during the transit," Concannon said. "The seas laid down for the first couple of days so that we could work. But they were building, and we knew they were going to be getting worse. So, we were racing the weather and racing the clock to get . . . something on board. Recover something so we could do something on the ship while we were shut down by the weather."

On March 6, the same day the storm smashed the fishing boat off the coast of Maryland a few hundred miles to the northwest, the crew of the *Seabed Worker* pulled up the first engine they had come across—though it wasn't clear which Apollo mission it was from.

Then the storm shut them down. All they could do was wait and hope that it would soon be over.

SATURN RAGED OVER them for five days, forcing the crew to keep the rovers out of the water. On the radar, the storm seemed to engulf all of the East Coast and much of the Atlantic, where the waves swelled and rocked the ship. The motion sickness scopolamine patches "were being distributed freely, that was for sure," said Capone.

They thought about trying to escape farther south to escape the storm. "The problem was the storm was so big that we couldn't get away from it," Concannon explained. There was no choice, so they "hung out and weathered it."

For those who weren't laid up sick, there was a darts tournament, which Bezos's father won. They caught up on sleep. And they lamented violating some of sailors' unwritten rules—especially the bananas.

"Mariners all over the world have a number of superstitions," Bezos said. "You're not supposed to ring the ship's bell. You're not supposed to bring backpacks onboard a ship, which unfortunately we did. We've got lots of bananas, and you're not supposed to bring bananas on board either. We are in the Bermuda Triangle. Somehow we now have the longest weather hold this experienced crew has ever seen."

But most of all they used the time to research the artifacts they had pulled aboard, and the ones that lay down some 3 miles below them.

"You don't just sit there and puke," Capone said. "You work. . . . Jeff put forth this vision and it was our challenge to make it happen. And I can tell you, none of the crew were daunted by the storm. Yeah, it was a pain in the neck. It wasn't comfortable. And this is a large ship, almost three-hundred-foot ship, getting battered around pretty good. But none of us was going to give up."

BY MARCH 11, the storm finally cleared, and the crew got back to work, deploying the rovers to the ocean floor. Soon they were hauling all sorts of metal aboard, working day and night. Bezos, unshaven, donned an orange jumpsuit, a hard hat, and protective goggles, and helped clear the mud from the engines.

The long hours, the isolation of the sea, and the hardship from the storm stripped away status, as they adhered to a credo true in all expeditions: if you're not contributing, you're taking away.

"We were hosing off artifacts and cleaning mud. We were all working," Concannon said. "We didn't have the CEO of Amazon. We just had Jeff and his mom [and dad] and his brother and his brother-in-law. . . . And everybody got their hands dirty. Everybody got little sleep. Everybody played darts when we were shut down by the weather."

After several days, they had a treasure trove of engine parts—enough to declare the mission a success and head home. They had chosen to bring the engines back to Port Canaveral, returning them to the place where they had blasted off more than four decades before. After three weeks on the ocean, the *Seabed Worker* approached the coast shortly after sunrise. The crew gathered on the deck and could make out Pad 39A in the distance.

They suspected that at least some of the engine parts they had recovered belonged to Apollo 11. But they didn't know for sure.

THE PARTS WERE brought to the Kansas Cosmosphere and Space Center, a museum that had worked for years with the Smithsonian Institution to restore and conserve aerospace artifacts. The staff there kept the parts constantly wet to prevent further corrosion, while gently rinsing the objects, blasting them with dry ice, and even removing sediment with dental picks. But they did not find any of the serial numbers that would identify what engine parts went with what mission. Based on the search area, Bezos's team was sure they had an engine from Apollo 11. They just needed to prove it.

Finally, one of the conservationists had an idea. Perhaps the serial numbers they were searching for were invisible to the naked eye, but would show up under a black light.

On his way to work one morning, he stopped to buy a black light and goggles. Later, as he shined it over a thrust chamber, he eventually saw "2044." He was so excited that as he ran to the phone to report his discovery, he tripped and fell, then scrambled back to his feet.

Bezos announced the news to the world in another blog post on July 19, 2013:

"When we stepped off the Seabed Worker four months ago in Port Canaveral, we had enough major components to fashion displays of two flown F-1 engines. We brought back thrust chambers, gas generators, injectors, heat exchangers, turbines, fuel manifolds and dozens of other artifacts—all simply gorgeous and a striking testament to the Apollo program. There was one secret that the ocean didn't give up easily: mission identification. The components' fiery end and heavy corrosion from 43 years under water removed or covered up most of the original serial numbers. We left Florida knowing the conservation team had their work cut out for them, and we've kept our fingers crossed ever since.

"Today, I'm thrilled to share some exciting news. One of the conservators who was scanning the objects with a black light and a special lens filter had made a breakthrough discovery—2044—stenciled in black paint on the side of one of the massive thrust chambers. 2044 is the Rocketdyne serial number that correlates to NASA number 6044, which is the series number for F-1 engine #5 from Apollo 11. The intrepid conservator kept digging for more evidence, and after removing more corrosion at the base of the same thrust chamber, he found it—'Unit No 2044'—stamped into the metal surface."

At a depth of nearly 3 miles, some 450 miles off the Florida coast, they had found the center engine of the rocket that first took men to the moon.

SINCE ITS FOUNDING in 1904, the Explorers Club had been celebrating the truly adventurous, and have counted some of the world's most courageous explorers as members, from Adm. Robert Peary and Matthew Henson, the first to reach the North Pole, to Ronald Amundsen, the first to reach the South. Charles Lindbergh was a member, as was Sir Edmund Hillary, the first to summit Mount Everest with his Sherpa, Tenzing Norgay. And, of course, it also celebrated those who pioneered space, including the crew of Apollo 11—Neil Armstrong, Buzz Aldrin, and Michael Collins.

Every year, the club threw a lavish, black-tie awards banquet at the Waldorf Astoria in New York, where the cuisine was as adventurous as the expeditions its members routinely undertook. On the menu were such delicacies as Earthworm Stir Fry, Maggot-and-Bug-Covered Strawberries, Scorpions on Toast, Duck Tongues on Belgian Endives, and Sweet and Sour Bovine Penis.

One year, the club's president made an entrance by riding a white horse onto the stage. The horse proceeded to poop onto the dinner plate of Edmund Hillary, who was seated on the dais.

For the 2014 dinner, on March 15, there was no horse. But there was the usual exotic fare: skewered cockroaches, North American beaver, an ostrich egg, tarantulas, goat and goat penis, and a pair of alligators, heads still attached, that were carved up before the guests like a pig.

After the feast, Buzz Aldrin came to the stage to introduce Bezos, who was accepting an award on behalf of the F-1 recovery team.

"Would you believe those rocket engines, those big F-1 engines that took us to fulfill the dreams of centuries and centuries," Aldrin said, apparently in awe. "The center engine of the first stage. Can you imagine—of all these engines, Jeff happens to find that particular engine."

The famous astronaut ribbed Bezos a bit about how secretive he and his space company had been.

"Jeff is trying to get people up to space," Aldrin said. "But he's not telling anyone about it. He's keeping pretty quiet. But I think he's going to tell me a little bit about it."

Bezos arrived on stage beaming, and joked, "I'm still making sure there's no cockroach in my teeth." In a short speech, he said that the team had "the feeling that we were recapturing history and making some history at the same time. I can tell you for sure we had a lot of fun doing it. It wasn't quick. It was difficult to find the engines, and the hard work of finding them with the side-scan sonar had to be done very painstakingly over a very large search area.

"I was blown away by the professionalism and the skill of the entire team. This isn't something that can be done by a small group. You need a bunch of professionals. The group who located the engines, they were just amazing. The ROV pilots—these guys are surgeons except they're working three miles down. The crane operators. Have you ever seen a crane operator work in heavy seas? The pitching deck. The whole crane starts to swing like a pendulum. These guys are such pros and it was a joy to watch every single person on our mission. This award is definitely accepted on behalf of the entire team."

He asked the crew to stand, urged the guests to applaud them, and then gushed, "These guys totally crushed it!"

THE F-1 TEAM wasn't the only space-related award of the evening. The president of the Explorers Club had become fascinated with Elon Musk, and what he was accomplishing in space, and had decided to present him with a special President's Award. As an image of Musk in a tight-fitting T-shirt, arms crossed, biceps bulging, showed on a huge screen, Musk hopped onto the stage to accept his award.

Like Bezos, Musk had come to the conclusion that the way to make spaceflight affordable was going to be re-creating a reusable rocket, one that could fly as frequently as an airplane. Only then would there be the big breakthrough that would allow space to open up to the masses—and allow him to get to Mars. SpaceX was getting close, and, unlike Bezos, who would never have discussed Blue Origin's plans so publicly, Musk used the speech to update the guests on the company's progress.

"I think what we've done thus far is evolutionary but not revolutionary," he began. "And the thing that needs to be accomplished by SpaceX or by somebody else is a fully reusable rocket system. That's the thing that really prevents us from establishing life on Mars.

"The way rockets work right now is they are all expendable. So, you fly them once, and you throw it away. You can imagine if any mode of transport was expendable, it wouldn't be used very much. But whether it's a plane, a boat, a car, a bicycle, or a horse—they're all reusable. If a 747 costs about a quarter-billion dollars and you need two for a round-trip, nobody is paying half a billion dollars from London to New York and back."

SpaceX had been working on developing the technology, and, as Musk said, "we're starting to do the bit where we bring the booster back." The rocket to be used in the company's next launch would be outfitted, for the first time, with landing legs. SpaceX would try to land it on a ship at sea first—not on land—"because we're not a hundred percent sure we can land it with pinpoint accuracy," he said. "So, coming back to land would be a dodgy prospect. But it's going to try to land to a fairly precise location, deploy the gear and then get recovered with a barge."

WHILE MUSK WAS talking publicly about how SpaceX planned to launch, land, and reuse rockets, Bezos was quietly laying out his plans as well—in an application with the federal government that had largely escaped notice.

On March 25, 2014—ten days after Musk detailed SpaceX's plans at the Explorers Club awards dinner—the United States Patent Office approved patent no. 8,678,321, titled "Sea Landing of Space Launch Vehicles and Associated Systems and Methods."

The ten-page patent laid out a system for recovering rockets that mimicked the approach that Musk had detailed in his speech. It laid claim to a system in which "a reusable space launch vehicle is launched from a coastal launch site in a trajectory over water." After the first stage's engines cut off and it separates and begins to fall back down, "the booster reenters Earth's atmosphere in a tail-first orientation. The booster engines are then restarted and the booster stage performs a vertical powered landing on the deck of a sea-going platform."

The technology was necessary, the patent explained, to bring down the costs of spaceflight and allow industry to push further more efficiently.

"Despite the rapid advances in manned and unmanned space flight, delivering astronauts, satellites and other payloads to space continues to be an expensive proposition. One reason for this is that most conventional launch vehicles are only used once, and hence are referred to as 'expendable launch vehicles' or 'ELVs.' The advantages of reusable launch vehicles (RLVs) include the potential of providing low-cost access to space."

The vision in the patent was expansive, detailing plans to land rockets not just in the ocean but on "other bodies of water including, for example, a lake, a gulf, ocean, sound or possibly even a large river." It covered the ability to launch rockets not just from land but "from sea on an ocean-going platform." And it discussed how the rockets would be serviced quickly, sometimes while on the barge being shipped back to sea, or how the booster would be transferred to smaller, faster ships in an effort to get them back to land quickly.

When Musk found out about it, he was furious. After the dispute over the rights to Launch Pad 39A, the patent was yet another indignity from what he felt was an inferior competitor. The idea of landing on ships at sea "is something that's been discussed for, like, half a century," Musk recalled. "The idea is not like unique. It's in fictional movies; it's in multiple proposals; there's so much prior art, it's crazy. So, trying to patent something that people have been discussing for half a century is obviously ridiculous."

It was years later, but the dispute still stayed with Musk.

"Jeff 'One-Click' Bezos," he said, referring to another of Bezos's controversial patents. "I mean, come on, Jeff. Leave it alone."

SpaceX promptly filed suit, challenging the patent. The idea for landing rockets on ships wasn't Blue Origin's invention, Musk's lawyers argued, it had been around for years, even if Blue Origin's patent only paid "lip service" to the existence of prior art.

If Blue's patent went unchallenged, it would hold exclusive claim over the ability to land rockets on ships—a potentially devastating blow to SpaceX. Like the Wright Brothers, who had won a sweeping patent for their flying technology in 1904, Blue would be able to freeze everyone else out—or demand licensing fees for the technology.

In its challenge, SpaceX demonstrated that others had conceived of the idea long before Blue Origin's patent—and had drawings to prove it. There was even a Russian sci-fi movie from 1959 that showed a rocket landing on a ship at sea.

Blue Origin withdrew the majority of its claims—a win for SpaceX. The ultimate victory, though, would come when the first rockets started landing.

DURING HIS SPEECH at the Explorers Club, Musk predicted that SpaceX would be able to pull off the feat of landing on a ship at sea in one of its upcoming launches. The odds weren't going to be great for its first attempt—"I think we have probably a forty percent chance of making that work," he said.

But the company would keep at it, like a gymnast working on his or her dismount, getting better with each attempt. "We've got a bunch of launches later this year, where the probability increases with each of those," he continued. "But I'm starting to feel a little more confident that this could work."

That, in turn, would help lower "the cost of moving to Mars to under a half a million dollars," a figure that he thought those who wanted to move to Mars could afford.

At the end of the evening, the president of the Explorers Club asked all the award winners and the presenters to come to the stage and be recognized once more. They came up one by one, more than a dozen in all. Musk was at one end of the stage. Bezos was at the other. They did not speak.

12

"Space Is Hard"

T HE ROCKET DETONATED suddenly into a giant orange fireball, sending a mushroom cloud floating ominously in an
otherwise pristine, sun-drenched Texas sky. Bits of debris scattered, trailing smoke and fire on the way down like fireworks in a display
as beautiful as it was violent.

As it climbed a few hundred feet over the company's McGregor test
site, the rocket had suddenly spun out of control and started falling back
down. Before it could get too far off course, its "flight termination system"
kicked in, blowing the rocket up a few hundred feet above the prairie. No
one was injured. And it was only a test, one that SpaceX stressed "was
particularly complex, pushing the limits of the vehicle further than any
previous test." Elon Musk even had coined an acronym for such spectacular failures: RUD, or rapid unscheduled disassembly.

But it was also a reminder that for all the advancements in rocket science, launches were really just controlled explosions of a combustible mix
of propellants. As Musk knew as well as anyone, one small error, even
something as small as a corroded nut, could send the whole thing up in
flames. "Still so damn intense," Musk had tweeted after a recent launch.
"Looking fwd to it feeling normal one day."

After the rocket test explosion, he tweeted: "Rockets are tricky."

Still, SpaceX had rattled off an improbable string of successful launches
of the Falcon 9 without a single failure. It was an amazing streak stretching over four years that was beginning to make the exercise feel routine.

But Musk still sweated every one, shooting an e-mail to the entire company, urging employees to step forward if anyone had a reason to call off the flight. He was the CEO-cum-wedding officiant—speak now or forever hold your peace.

SpaceX's success had raised expectations, but its brashness had attracted criticism. Its fan base was huge and growing. The SpaceX page at Reddit, the social media site, had ten thousand subscribers in June 2014. More and more people were buying the $22 "Occupy Mars" T-shirts from the SpaceX online store. And Musk had become something more than a business executive; he was now a cult figure, whose legend was growing well beyond Silicon Valley.

With Tesla and Solar City, the solar energy company, he had set out to transform American transportation and energy use. SpaceX was an improbable success story that was not only disrupting the industry but singlehandedly reigniting interest in space. CBS's *60 Minutes* declared that Musk had built an "industrial empire." *Time* magazine put him on the cover of its "100 Most Influential People" issue. The *Atlantic* canonized him as "perhaps this era's most ambitious innovator."

"In the spirit of inveterate and wide-ranging tinkerers like Leonardo da Vinci and Benjamin Franklin, Musk has transformed virtually every field he's taken an interest in, from electronic payments to commercial spaceflight to electric cars," the magazine wrote. "The range and scale of Musk's ambitions have attracted skepticism, but over time, he has proved himself to be not only an ideas man but an astute business thinker."

But within the somewhat clubby space community, SpaceX was becoming the company people loved to hate. At a space industry party, there was a photo Musk taped to the inside of the toilet so that his competitors could take turns pissing on him.

Once dismissed as an "ankle biter," SpaceX was now a formidable competitor, one to reckon with. It was also gunning for the United Launch Alliance's breadbasket—the lucrative launches for the Pentagon and intelligence agencies.

For a decade, the company had a monopoly on the contracts, worth hundreds of millions of dollars. A decade earlier, Musk had sued, arguing that SpaceX should be allowed to compete. But without a rocket capable of flying, the suit was dismissed.

Now he had a rocket. He just didn't yet have the US Air Force's certification required for the launch—and the Pentagon was about to award another big batch of contracts to the Alliance, effectively locking SpaceX out for years. Filing another lawsuit would be risky—it's usually not the best business practice to sue the agency you're trying desperately to get to hire you.

The list of cons was long. But the pros were substantial as well. National security launches paid big money—the multiyear program could be worth as much as $70 billion—and SpaceX knew it could undercut Lockheed's and Boeing's prices, disrupting the market, giving it a stream of revenue that could sustain it for years and help it get to Mars. But the clock was ticking. If the company was going to protest the contract, it had to act quickly.

"Suing the military industrial complex is something you do not take lightly," Musk recalled.

During a visit to Washington, DC, while sitting in the back of a sedan after a speech, a pair of his advisors asked him what he wanted to do.

Musk went quiet, closed his eyes, and put his head back. He stayed that way for two minutes, then three. A long time. He had several quirks, and his sudden retreat into his own mind was an eccentricity the people at SpaceX were used to. People coming in to interview with Musk were sometimes warned that when he goes silent, it was because he's thinking and it's best not to interrupt him. The advisors knew not to say a word. Six minutes passed. Then eight. An eternity.

"I'd seen him go Zen before, but I'd never seen him go this Zen," one of the advisors recalled.

Then Musk opened his eyes. "File the lawsuit," he said. He got out of the car and went to the next event.

The advisors looked at each other and one of them said, "He just teleported himself into the future!"

ONCE THE LAWSUIT was filed, Musk continued to attack throughout the spring and summer of 2014, delighting the press corps in the nation's capital, who were unaccustomed to such a bombastic character. "Musk is," one defense reporter wrote, "a good interview."

Which meant he was unfiltered, a refreshing change of tone in a buttoned-up town where officials rarely deviated from the script. At an event at the National Press Club, Musk defended the lawsuit, saying SpaceX should be given the opportunity to compete, and he derided the air force's certification process, calling it a "paperwork exercise." If his rockets were good enough for NASA, he said, they should be good enough for the Pentagon.

He purposefully picked a fight with the United Launch Alliance (ULA). It was the dominant player, but it had a big weakness. The RD-180 engines it used for its Atlas V rocket were made in Russia, and this was coming at a time of increased tension between the United States and Russia over the latter's annexation of Crimea. Musk went after the Alliance relentlessly.

At a reception at the Newseum near the Capitol that spring, as he showed off the crew version of Dragon, Musk stood in the middle of a massive media scrum to deliver an indictment of the entire process.

"Our toughest competitor on the international launch market is the Russians, and the US Air Force sends them hundreds of millions of dollars every year for Russian engines," he said. "It's super messed up. I mean, what the fuck, you know? . . .

"Can you imagine if you went back forty years ago and told people that in 2014 the United States would be at the mercy of Russia for access to low Earth orbit, let alone the moon or anything else? People would have thought you were insane. It's just incredible that we're in this position. Something needs to be done to get us out of this."

In response to a question about the wisdom of going up against the Alliance, he said that "Eisenhower warned about the military-industrial complex, and he ought to know. Has it gotten better or worse since Eisenhower? It hasn't gotten better. . . . Lockheed and Boeing are used to stomping on new companies, and they certainly tried to stomp on us. I think we've got a shot at prevailing. We're certainly a small up-and-comer going against giants."

The bombast, the lawsuit, the media attention began to tick off the Pentagon. The head of the air force's space command at the time told a reporter, "Generally, the person you are doing business with you don't sue."

For the first time, the Alliance began to fight back publicly, highlighting its long history over SpaceX's inexperience in a marketing campaign—what it called "results over rhetoric."

"The whole tenor of the campaign is to make perfectly clear that there is a lot at stake when it comes to successful space launches—literally lives are at stake," Mike Gass, the CEO of the Alliance, said at a press conference. "We also want to make clear that there is a big distinction between a company that has a hundred-year combined heritage in successfully delivering satellites into orbit and a company that is not yet even certified to conduct one [national security] launch. . . .

"SpaceX is trying to cut corners and just wants the USAF to rubber stamp it," Gass said. "SpaceX's view is just 'trust us.' We obviously think that's a dangerous approach and, thankfully, so do most people."

SpaceX didn't have anything to lose, and was ready to get into a fight.

"ULA doesn't believe in competition. Monopolists never do," spokesman John Taylor said in a statement. "In ULA's case, it would rather call a press conference to announce an inside-the-Beltway lobbying campaign aimed at distracting lawmakers from the benefits competition brings to the marketplace: better technology, improved reliability and affordable prices."

Two months later, Gass was ousted. Tory Bruno, the Alliance's new chief executive, was brought in to make the company leaner, more efficient—able to compete with SpaceX, which was now threatening its business. Bruno vowed to "literally transform the company" by cutting the price of launch in half and developing a new rocket.

In addition to streamlining the business, the Alliance had a secret weapon in the war against SpaceX: Jeff Bezos.

FOR YEARS, BLUE Origin had been building a monster of a new rocket engine, one that stood 12 feet tall and had 550,000 pounds of thrust—even more than the engines that powered the space shuttle. The BE-4, as it was known, was not as robust as Bezos's beloved F-1s, the most powerful rocket engines ever built. But it was designed to be a reliable workhorse, one that could fly again and again at a relatively low cost.

The fact that Blue Origin was developing its own engine, and building the infrastructure in West Texas to test it, was yet another sign that Bezos was dead serious about space, and that he had poured a vast amount of resources—perhaps even as much as $1 billion—into the development of the engine alone.

At a press conference at the National Press Club, Bruno and Bezos sat side by side before a banner with the phrase "Igniting the Future" to announce they were joining forces—Blue Origin would sell the BE-4 to the Alliance. That would allow the Alliance to avoid using the Russian-made RD-180—and just as important, take away Musk's line of attack.

It was a shocking and unlikely marriage—the Lockheed Martin–Boeing conglomerate, which together had a century of experience in space, with Blue Origin, the quiet upstart that had plodded along carefully in the shadows. But now for one of the first times, Blue Origin was standing squarely in the spotlight—and it was doing so with SpaceX's archenemy.

"It's kind of the best of both worlds," Bruno beamed. "We have their innovative, entrepreneurship together with ULA's solid track record of success, certainty and reliability."

Bezos praised his new partner and its long heritage, noting that the Alliance "has for the last eight years put a satellite into orbit almost once a month. It's an unmatched record of success and an incredible tribute to detail orientation and operational excellence."

He geeked out over the technical details of the engine, discussing how its "oxygen-enriched stage combustion cycle" was better than a "gas generator" and how the engine has only a "single turbo pump" and had just "one shaft, so it's as simple as it can be while still being high performing and highly reliable."

Later that day, when Musk was asked about the United Launch Alliance–Blue Origin partnership, he was, as always, blunt: "If all your competitors are banding together to attack you, that's, like, a good compliment," he said. "I think a very sincere compliment."

It also increased the pressure on SpaceX. Musk couldn't afford a misstep. Not now. Not with his rivals gunning for him, and the Obama administration investing heavily in SpaceX, and Musk now a celebrity with the ability to move markets and make the media swoon with a single tweet. All of it was building to a Hollywood-like crescendo that was propelling Musk and his space company higher and higher, to a rarefied altitude where it finally had something to lose.

SEPTEMBER 16, 2014, was the 1,167th day since NASA had launched an astronaut from American soil, an embarrassing streak that stretched back to the last shuttle flight in 2011. Every day without a crewed launch brought NASA closer to breaking an ignominious record: the 2,098-day hiatus in human spaceflight between the last of the Apollo launches and the first shuttle flight in 1981.

But on the 1,167th day, the space agency had good news: its plan of how it would fly astronauts once again, an announcement that NASA administrator Charlie Bolden said set "the stage for what promises to be the most ambitious and exciting chapter in the history of NASA and human spaceflight."

Two companies, SpaceX and Boeing, had won the contracts as part of NASA's "commercial crew" program to fly the next generation of astronauts to the International Space Station, the agency announced. The companies would fly the same number of flights, and be required to hit the same milestones. But SpaceX had simply bid less, and as a result laid out in stark contrast the difference between itself and its rivals.

Boeing's award was $4.2 billion. SpaceX would receive $2.6 billion.

Musk had been saying for years that SpaceX could fly cheaper and more efficiently than the traditional contractors, and NASA was taking him up on it—while also hiring Boeing, the more expensive, and experienced, company.

By now, SpaceX had flown the Falcon 9 and the Dragon spacecraft to the station multiple times. But Musk was eager to move to the next phase in his quest to get to Mars—flying actual people in his new version of the Dragon spacecraft. It looked like a sleeker, sexier version of the capsules that had taken the Apollo astronauts to space. Outfitted with reclined seats, giant screens, and a shiny interior, it could have passed for the VIP, bottle-service section of a nightclub. (In addition to being SpaceX's founder and chief executive officer, he had the title of lead designer.) But unlike traditional spacecraft that splashed under parachutes into the ocean, the Dragon had its own engines, giving it the ability to land propulsively—using engine thrust to slow itself down—virtually anywhere on Earth.

"That is how a 21st century spaceship should land," he said.

The White House's risky bet to rely on the commercial sector, now being led by SpaceX, seemed to be playing out just as the Obama administration had hoped.

Four days after winning the commercial crew contract, SpaceX had flown yet another successful cargo flight to the station, and the Dragon was about to return home. Orbital Sciences, which along with SpaceX had been hired by NASA to fly passenger-less cargo missions to the station—had launched its Cygnus spacecraft to the station three times.

And now, on October 28, they were about to fly again. Leading up to the launch, Frank Culbertson, Orbital's executive vice president and a former NASA astronaut, joked that the astronauts aboard the station might need "some of those red and green wands they use on the deck of an aircraft carrier" to direct all the spacecraft traffic coming and going.

Like SpaceX, Orbital was ready to fly to the station "more and more frequently and do it for many years to come," he said. The goal was to make access to the station routine—"a stepping stone to what we're going to do next, which is to go beyond low Earth orbit, go out to the moon, continue to explore that, and eventually go to Mars and to asteroids and to continue to explore our solar system."

The weather was perfect for the launch, and the mood was upbeat. NASA—and the commercial sector—had momentum now and were eager to keep it going, even if some were concerned about allowing the still-young industry to fly NASA's most precious resource: its astronauts.

FOR THE EVENING launch, crowds gathered along the Virginia shoreline, several people deep. Kids sat on parents' shoulders for a better view. Some even crept on top of cars. The air was full of the illuminated screens of cell phone cameras ready to record the rocket blasting off. They counted down in unison: "Five. Four. Three. Two. One!" And cheered as Orbital's Antares rocket lifted off over a ball of yellow-orange fire and smoke at 6:22 p.m., 15 minutes after sunset.

But within seconds, the majesty of the liftoff morphed into a menacing flash as the rocket exploded into a shrapnel-spewing fireball. A massive mushroom cloud filled the sky, scattering bits of debris like fireworks. A few miles away, the spectators could see it—and feel the heat—before they could hear it. Then, the blast arrived like a cannonball, knocking some off their feet and sending others running for cover.

The explosion incinerated the rocket and the 5,000 pounds of cargo it was to ferry to the International Space Station. It devastated the launch-pad, leaving a crater 30 feet deep and 60 feet across that would cost $15 million to repair.

The hole it left in NASA's plans to rely so heavily on the commercial sector, however, was even bigger.

THREE DAYS LATER, on October 31, 2014, Branson was home on Necker Island, his retreat in the Caribbean, talking on the phone to his son, Sam, who had just completed training in a centrifuge near Philadelphia, preparing for his spaceflight.

For years, Branson had been promising that the "world's first spaceline" would soon be flying tourists to space. The first flights were supposed to start in 2009, but the date slipped again and again, until the company stopped providing dates. Instead it said that while it "expects to be the first company to provide sub-orbital flights to the general public (and certainly the best!)" it would launch only "when we are happy with the results of the exhaustive WhiteKnightTwo and SpaceShipTwo flight programme."

It didn't, however, tone down its promotional materials. For $250,000, Virgin promised quite a thrill: "Astronauts tell us that nothing can really prepare you for your first experience of space, but we will ensure that you are fully equipped to savour every second of an experience which will be intense, wonderful and truly unforgettable," it declared.

Now, after years of delays, Virgin Galactic was getting ready to finally fly. Although company executives had warned him not to reveal a timetable for flights, Branson couldn't contain himself. The company was so close that he'd told the media the first human test flights to space would take place by Christmas. Then, he and his son would go together in the early part of 2015, followed by paying customers.

The spacecraft had completed more than fifty test flights, but most of them were just unpowered "glide flights." It had tested the "feather system," the Rutan invention that would help it gently return to Earth, ten times. But even though it had conducted only four powered test flights by firing the engine, Branson and Virgin Galactic had gone into full marketing mode.

Virgin Galactic had signed up sponsorships with Grey Goose, the vodka maker, and Land Rover, which had sponsored a contest to send four winners to space. It had inked a deal with NBC to broadcast the first flight "in a primetime special airing on NBC the night before the launch, and a 3-hour live event on 'TODAY,' hosted by Matt Lauer and Savannah Guthrie," the companies said in a press release. And it had signed a deal to fly its paying customers out of Spaceport America, the futuristic spaceport in New Mexico that cost taxpayers $220 million to build.

The first flights were nearly here—just weeks away—and Branson was already looking ahead to the future.

"I think we can in the years to come bring the price down so that a lot of people will have the chance to become astronauts," he told an interviewer.

Like his son, Branson had also been training to go to space. Less than two weeks earlier, he had taken an acrobatic flight in a stunt plane to get his body used to the additional g-forces.

Flying above the Mojave Desert, he was ever the showman, asking his pilot, "Are we able to do a loop? Why don't we do a little show-off over the runway."

"You feel okay?" the pilot asked as they spun upside down, Earth swirling beneath them.

"Absolutely perfect," Branson enthused. "That was brilliant!"

But now at Necker Island, he was talking with his son about his experience in the centrifuge, when he got an urgent note. It was from George Whitesides, the CEO of Virgin Galactic. There had been a catastrophic accident. Branson had to go.

PETER SIEBOLD WAS in the cockpit again. The forty-three-year-old test pilot for Burt Rutan's Scaled Composites, which had built and designed SpaceShipTwo for Virgin Galactic, was ready to fly this time.

A decade earlier, during the Ansari X Prize, he'd had misgivings about the safety of SpaceShipOne, and dropped out of the program. But he had stuck with the SpaceShipTwo development, and on the morning of October 31 strapped himself into the copilot's seat, alongside Michael Alsbury, the other test pilot.

They were close friends whose children played together on weekends. Both were self-taught flying obsessives who wanted their pilot's licenses far more than their driver's permits. They had gone to the same university, and jumped at the chance to become a test pilot for Rutan, trying out his latest inventions. Alsbury had been picked to fly the last vehicle Rutan built—a flying car—before he retired in 2011. Siebold had grown up flying in his father's plane, from being perched on a stack of pillows when he was five years old so that he could experience the thrill of flight as if he were the pilot.

Now it was a bright October morning, and SpaceShipTwo was attached to the belly of its mothership, WhiteKnightTwo, which was taking the spacecraft up higher and higher. When it came to 50,000 feet, it was released and the pilots ignited the engines.

Siebold had considered the mission "high risk." When asked why, he said they were "doing a significant envelope expansion that day. Flying an unproven rocket motor in an unexplored aerodynamic regime . . . classic test hazard assessments would categorize that as a high-risk flight."

Also, the crew was "using a propulsion system that history has shown can be unreliable, or much less reliable than a turbine or reciprocating engine."

Given how risky the flight was, Siebold took a moment to collect himself just before the spaceship was to be released from the mothership. He ran his hands over the rip cord of his parachute and his oxygen mask and his seatbelts, as if rehearsing the steps he would have to take in an emergency and to improve his "muscle memory."

After WhiteKnightTwo released the spacecraft, Siebold and Alsbury fired the engines and soon they were soaring toward the heavens, breaking the sound barrier, 10 miles high.

"Ignition! #SpaceShipTwo is flying under rocket power again. Stay tuned for updates," the company tweeted at 10:07 a.m.

The update six minutes later wasn't good: "#SpaceShipTwo has experienced an in-flight anomaly. Additional info and statement forthcoming."

THE LAST THING Siebold remembered of being in the spacecraft was a sickening jolt, grunting noises, a loud bang, and then cabin depressurization. The aircraft pitched up violently, he would tell investigators from the National Transportation Safety Board. The sound of the spacecraft coming apart was odd, delicate even, like "paper fluttering in the wind." Then the crushing g-forces caused cerebral hypoxia, a lack of oxygen to the brain, and he blacked out.

When he came to, Siebold was outside the spaceplane in a free fall, his helmet askew, his oxygen mask shifted. The wind howled in his ears. The extreme cold air engulfed his body. There was something bothering his eyes, and when he opened them, he saw the wide expanse of the desert below.

The Mojave Desert floor rising fast.

Just as he had rehearsed a few minutes before, Siebold used his hands, Braille-like, to find the buckles of his seatbelt and release them. Falling through wispy cirrus clouds, his training kicked in, and he went into the free-fall position, opening his arms and legs out spread-eagle-style, to create drag.

The next thing he remembered was another jolt that surprised and possibly woke him up. He told investigators he wasn't sure whether he had gone unconscious again. But if he had, the snap of the bright red parachute deploying automatically brought him back. His shoulder was killing him—he thought it was dislocated. As he floated under the parachute, he tried unsuccessfully to shove it back in place so that he could use it to steer.

He braced for a hard landing, right into a windswept creosote bush in the middle of the desert. As he waited for help, he noticed his chest was covered in blood. His arm was broken in four places and his right hand was numb, "as if he were out throwing snowballs without gloves." He had corneal scratches in his eyes and he later had a piece of fiberglass removed from his left eye when at the hospital. But he was alive.

Emergency crews would find Alsbury's lifeless body not far away in the wreckage, still in his seat. The coroner determined the cause of death was "blunt force trauma to the head, neck, chest, abdomen, pelvic area and all extremities and internal organs."

He was thirty-nine years old, with two children, aged ten and seven.

AFTER HANGING UP on the call with his son, Branson hopped in a plane and headed to the crash site. He knew he had to get there as soon as possible.

This was the second fatal accident for the program. In 2007, three employees of Scaled Composites had been killed during a ground test of the engine's nitrous oxide system. The explosion had charred the desert floor, which had looked like a war zone, with debris scattered everywhere and multiple people injured.

The California Occupational Safety and Health Administration fined Scaled Composites $28,870, later dropped to $18,560 after appeal.

The explosion was "obviously horrendous for the families and a big setback for our program as well," Branson recalled later. "And after that I think we decided to take the testing in house and have our own team do it."

That same year, his train company had had a fatal derailment in northeast England. He had gotten to the scene as soon as possible, racing over from a vacation in Zermatt, Switzerland.

"I knew the importance of getting there as fast as possible and being there as fast as possible and confronting it head on whether it's your fault or not—actually particularly if it's your fault," he recalled.

Once he arrived, he spoke to the Virgin Galactic team before talking to the media.

"I addressed them all and reassured them as best I could that they had built a beautiful craft," he said. "We had the biggest hug in history, and I made it clear we would continue knowing the spaceship was fundamentally fine."

But the investigation by the National Transportation Safety Board was only just beginning. And questions from the press were mounting.

On the *Today* show, three days after the crash, Matt Lauer pressed Branson on the future of the company.

"People are already wondering whether this accident and the death of this copilot is a crippling blow," Lauer said. "There have been delays; there have been setbacks in the past with this program. Can Virgin Galactic survive the image that has been seen around the world of that vehicle coming apart at forty-five thousand feet?"

Was it, Lauer wanted to know, worth the risks?

Branson had thought about this very question on the plane ride to the Mojave Desert. He'd been happy to risk his own life in hot-air balloons, in speedboats, and in all manner of stunts, that were daring and dangerous and good for business. But this was none of that. This was a sobering stop to a carousel that had perhaps spun too fast and too loudly for too long.

Maybe he should give it up. Maybe space was too hard. The company had spent $500 million or more on the venture and had still not flown a single person to space. But as Branson landed and met the team, they'd urged him on. He began to think he owed it not just to Alsbury who wouldn't have wanted them to quit, but to them. Hadn't every explorer

faced such difficulty? This was the moment Virgin Galactic and the industry had feared—and prepared for. This was their crucible, their Apollo
1 moment. The time to decide whether they would retreat or reassemble
and attack again stronger, if scarred.

"Absolutely, it's worth the risks," Branson said. "It's a grand program,
which has had a horrible setback. But I don't think anybody watching this
program would want us to abandon it at this stage."

He'd made his decision. They'd carry on.

UNTARNISHED AND SEEMINGLY invincible, SpaceX ended 2014, one
of its most successful years, with a massive holiday party. Why not celebrate? It had proved it could fly reliably to the International Space Station.
It had won the right to fly astronauts, which combined with the cargo
contract amounted to more than a $4.2 billion investment from NASA.
It was signing up new commercial satellite customers, had added 39A to
its mantle, was winning the war in Washington over the national security launches. And it was getting closer to landing a rocket safely so that
it could be reused.

It was on a roll.

The holiday party was so big that a map detailed the various venues—
including an indoor beach with hammocks and trucked-in sand, a casino, an "unclean room" where employees put on a full-body white suit
and painted away. There was a dance hall with swirling lights and circus-
like acrobats who swung from giant hula hoops from the ceiling. Rick-
shaws and a mini-train, the "SpaceXpress," ferried partygoers across the
great bacchanal, from the stocked bars to the game room with glowing
foosball table and the "Candy" room with treats to be nibbled from the
wall and a giant desert display. There was also a wall of donuts laid out in
the form of the SpaceX logo and an adult-size ball pit that was, of course,
right next to a bar.

"My favorite part of the @SpaceX X-Mas party—definitely the ball
pit!" tweeted Garrett Resiman, a SpaceX executive and former NASA astronaut, who flew on two space shuttle missions and spent three months
on the International Space Station.

SIX MONTHS LATER, they filed into SpaceX's headquarters around
dawn on a Sunday, ready to party again, ready to celebrate yet another

milestone. They packed in several deep around mission control for the launch of the Falcon 9 scheduled for just after seven a.m. Pacific time on June 28, 2015. It was a lovely morning in Florida: mid-80s, light wind. There was only a 1 percent possibility of calling off the launch because of weather.

The drama leading up to the flight wasn't the launch, but the landing. Or rather, the landing attempt. For months, SpaceX had been practicing an unprecedented maneuver, flying its first stages back to land on what it called its "autonomous spaceport drone ship." Although each of the previous two attempts had ended in fireballs—"rapid unscheduled diassamblies"—the company was getting very close.

On both tries, the rocket had actually hit the ship, an incredible achievement considering the booster had to fly back to Earth after screaming into space. But each time, something had gone wrong at the last minute, giving the company enough material for a fiery blooper reel.

Now, Musk felt that SpaceX had finally figured out how to do it, and was confident about its chances. He had invited a crew from National Geographic to film at SpaceX's headquarters to capture what would be a significant moment in the history of spaceflight.

If SpaceX could successfully land the first stage, it would be the first time anyone had done it—a huge leap forward for the company and the industry as a whole. And it would send a defiant message to the Alliance and Blue Origin. It was also Musk's forty-third birthday—what better way to celebrate than to make history, while defying your critics?

The launch, too, was critical.

Seven months after Orbital Sciences' rocket blew up on its way to deliver supplies to the space station, a Russian spacecraft, laden with thousands of pounds of cargo, supplies, and food, started spinning wildly, as if it were an out-of-control amusement park ride in zero gravity.

Another launch to the space station. Another failure.

Now, it was SpaceX's turn.

After the back-to-back failures, the increased competition from the Alliance and Blue Origin, and the heightened expectations that came with winning the award to fly astronauts to the station, the pressure was on for SpaceX. Another disaster would undermine the Obama administration's bold experiment to contract out missions to the space station so that NASA could pursue the more grandiose mission of flying to Mars.

And it would fuel questions about a still-nascent industry where failure is common, expensive, and measured in mushroom-cloud plumes of smoke.

Then there was the question of what two failed missions had cost the space station. In the summer of 2015, NASA officials maintained that the astronauts aboard the orbiting lab were in no danger. But a NASA slide showed that with current food levels, the space station would reach what NASA calls "reserve level" by the end of July and run out by September 5.

AT T-MINUS 13 minutes, the launch director conducted the go/no-go poll of the thirteen members of the launch readiness team, a last check of all the systems and stations to make sure the rocket was ready. It was a call-and-response ritual passed down from NASA, and at SpaceX they did it now with ease and confidence.

"All stations verify ready for launch," said the launch conductor into his headset, as he began the poll, surveying the teams in a call and response, each ending with an affirmative "Go!"

Propulsion was go. So was Avionics. It was a go from Guidance, Navigation and Control, and the chief engineer, go all the way down the line until the mission director said, "MD is go!" and the launch director said that the "LD is go to initiate terminal count."

THE COUNTDOWN WENT smoothly, and soon: "We have liftoff of the Falcon 9. The Falcon 9 has cleared the tower."

The employees gathered at SpaceX's headquarters cheered wildly, as if at a football game, pumping their fists at another successful liftoff. Ready to party once that thing landed.

As the rocket climbed, it looked excellent—or "nominal," which in rocketry means everything is going fine.

"Stage one propulsion is nominal," the propulsion engineer said shortly after liftoff. "Power and telemetry nominal," said the avionics engineer.

At T-plus 1 minute and 30 seconds, the Falcon 9 passed through what's known as Max-Q, or maximum dynamic pressure, the moment when the rocket is under the most stress as it screams skyward. Still, the status was good: "Stage one propulsion is still nominal."

At T-plus 2 minutes, the Falcon 9 was at an altitude just shy of 20 miles, racing to space at 0.6 mile per second. The fiery plume of smoke and fire behind it had expanded, which was normal given the reduced air pressure at that altitude. Everything was going smoothly.

Until it wasn't.

Just over two minutes after liftoff, the rocket exploded, engulfed by a white, wispy cloud. After a little while, the smoke and debris dissipated, leaving only the pale blue sky. It was as if a magician had somehow made the rocket, and the 4,000 pounds of cargo it was carrying, disappear.

On the space station, NASA astronaut Scott Kelly tweeted that he had watched the launch from the station. "Sadly, failed Space is hard."

The crowd at SpaceX headquarters went silent, some with their hands over their mouth. The National Geographic team kept the cameras rolling, capturing the devastating silence in what was supposed to be a birthday victory celebration but now felt more like a funeral.

A FEW WEEKS after the launch, SpaceX would identify the cause: a single faulty steel strut, 2 feet long and 1 inch wide at its thickest point. It was supposed to be able to withstand 10,000 pounds of pressure, but had buckled under 2,000, causing helium to overpressurize an oxygen tank in the second stage, which had led to the explosion.

In a call with reporters a month later, Musk was every bit the rocket scientist he had become, giving a preliminary but detailed analysis of the failure. But it was also as if he were delivering a business-school lecture on how a successful startup can retain its innovative culture and edge as it grows into a corporate behemoth.

He would point to another possible cause, saying that as the company continued to grow, it may have lost some of the inherent paranoia that fueled SpaceX in its early days, when it was unclear whether it would ever be able to launch rockets reliably.

After a string of successes, the explosion was the "first time we've had a failure in seven years" outside of the test flight, Musk said. "To some degree I think the company as a whole maybe became a little complacent."

When the company lost a string of rockets in its early days, only a few hundred people were working at SpaceX. Now, there were four thousand. "The vast majority of people at the company today have only ever seen success," he said. "You don't fear failure as much."

And so, when he sent out his wedding officiant e-mail before each launch, asking people to come forward, it didn't "resonate with the same force" as it had when the company was small and scrappy and feared going out of business.

The reaction became "There's Elon being paranoid again," he said.

But now even the uninitiated knew the driving power of failure—and fear—"and we'll be the stronger for it," he said.

13

"The Eagle Has Landed"

F OR YEARS, JOURNALISTS had been banging at Blue Origin's door, trying to get a glimpse into a mysterious company that operated like the Central Intelligence Agency. Now, on the morning of November 24, 2015, Blue Origin was reaching out to them, ringing their cell phones in the predawn darkness. The groggy journalists were told to check their e-mail for a press release that just went out, and that they'd be assigned a slot to speak with Bezos later that day.

He had news to share.

The day before, Blue Origin had launched a rocket deep in the West Texas desert that traveled past the threshold of space, hitting a top speed of Mach 3.72. New Shepard, the suborbital vehicle named for Alan Shepard, the first American in space, had flown to 329,839 feet, or 62 miles, past the 100-kilometer "Kármán line" that's widely considered the edge of space.

The capsule on top of the rocket, which had no passengers in it, separated from the booster and landed softly under the guidance of parachutes. More important, the rocket landed after falling back and enduring 119 mph high-altitude crosswinds. Using a GPS guidance system, and a fin system that helped stabilize it on the descent, the booster fired its engine to slow itself down before deploying its landing legs and touching down softly on a concrete landing pad.

It hit 4½ feet from the center. For a first landing, that was a bull's-eye.

At the Blue Origin headquarters, employees had gathered together to watch the landing on television. And as the rocket stood there, pandemonium broke out, some four hundred engineers cheering wildly, pumping fists and hugging one another.

From the beginning, Blue Origin had been trying to build a reusable rocket, one that could be launched, and then fly again, like an airplane, a breakthrough the industry had long been waiting for. It would, at last, lower the cost of space travel, and make it accessible to the masses. Now Blue had pulled off the landing, a triumphant crescendo of more than a decade of work.

Bezos was beaming. In interviews afterward, he called it a "flawless" mission and "one of the greatest moments of my life. I was misty-eyed."

He had founded the company fifteen years earlier, and had decided to build a chemically fueled rocket that would be reusable a few years after that. Now, Blue Origin had finally done it.

Later, he said that the joy the landing gave him reminded him of the saying "God knows how to appropriately price his goods."

"The things that you work hardest for, for the longest periods of time, always bring you the most satisfaction," he explained. "If you do something and it takes you ten minutes to do how satisfying can it actually be? You work on something for a decade and it finally comes out the pipeline. And for me, in a sense I've been working on that since I was five, so it was incredibly satisfying. And I think the whole team felt that. The people who go into this business do it because they are missionaries."

As it stood on the landing pad, charred from the flames, the rocket stood as a testament to math, engineering, and science. It was unlike any rocket that had ever flown before.

Traditional rockets were all brawn and no brain, powerful boosters with a single job: to wrest their way out of gravity's grip. Once they had done that, they were expendable, falling into a watery grave.

But the New Shepard was brains and brawn, an autonomous robot that could fly itself. Guided by computer algorithms, sensors that measured the wind speed, and a GPS system, it fell back to Earth until its engine refired at 4,896 feet above the ground, slowing it down on its approach to the landing site.

There the most remarkable part of the descent happened. For a short moment, the rocket hovered over the pad, taking a moment to check the coordinates to make sure it was in the right spot. Its system decided it wasn't, so it used its thrusters to nudge itself over just ever so slightly, a maneuver that caused the New Shepard to sway over and then back, as if it were scooching over a seat on a couch. Once it was apparently pleased with its location, it touched down in a plume of dust and smoke, kissing the pad gently at 4.4 mph.

This would put Blue Origin on the path to its first goal of flying paying tourists past the threshold of space, allowing them to enjoy the view from above, the curvature of Earth, the thin line of the atmosphere, the vast darkness of space beyond. For this flight, the company had also tested the crew capsule, without passengers in it. It landed as well, under parachutes, eleven minutes after lifting off.

Bezos said the flight was also a significant step toward its longer-term goals of building an even more powerful rocket, which at the time they were calling only "Very Big Brother." In his taunt two years earlier, Musk had said that the chances "of unicorns dancing in the flame duct" were greater than Blue Origin's building a rocket capable of delivering a payload to orbit. But now, here was Bezos saying they were doing exactly that.

It would launch in full view of SpaceX's Pad 39A. A couple months earlier, Bezos had announced that Blue Origin was taking over Launch Complex 36. While its pedigree was not as rich as that of 39A, Pad 36, located just down the road at Cape Canaveral, had been used for forty-three years before it was shuttered. It was home to 145 launches, including the Mariner missions, designed to be the first US spacecraft to fly by other planets, such as Venus and Mars. Pioneer 10, the first spacecraft to travel through the asteroid belt, also launched from there.

But like much of the Florida Space Coast's infrastructure, it had been abandoned and was rusting away. "The pad has stood silent for more than 10 years—too long," Bezos said at the unveiling ceremony. "We can't wait to fix that."

Now, with the landing of New Shepard, he had another victory to celebrate. And he took to Twitter—Musk's preferred medium—to announce the endeavor to the public:

"The rarest of beasts—a used rocket," Bezos wrote in his first ever tweet, even though he had joined Twitter in July 2008. "Controlled landing not easy but done right can look easy."

To Musk, the level of celebration surpassed the height of the accomplishment. And now, after the fight over 39A, the patent dispute, the teaming up with the Lockheed-Boeing alliance, and the tensions of employee poaching, he was fuming.

Bezos's celebration was not only unseemly showboating, Musk thought, but factually inaccurate.

Years before, SpaceX had repeatedly flown a test rocket called Grasshopper a few hundred feet into the air and then landed it, with one flight as high as nearly half a mile. So, technically, Musk had done it first.

"@JeffBezos Not quite 'rarest'. SpaceX Grasshopper rocket did 6 suborbital flights 3 years ago & is still around," Musk tweeted in response. He added, "Jeff maybe unaware SpaceX suborbital VTOL [vertical takeoff and landing] flight began 2013."

But the highest any of those test rockets had traveled was 1,000 meters (3,280 feet). New Shepard's rocket hit an apogee of 329,000 feet and its capsule went even higher. No rocket had ever made it to space before and then landed vertically. That was a first—and a record for the history books.

What also bothered Musk was that the general public didn't seem to understand the difference between what SpaceX was doing and what such companies as Blue Origin and Virgin Galactic were attempting. SpaceX's rockets were launched into orbit; theirs went only to suborbital space and then came back down.

For years, Musk had tried to make interviewers, and the general public, understand the distinction. He even had SpaceX's press people call reporters to impress the difference upon them. Reaching the threshold of space in a simple up-and-down endeavor was—"like shooting a cannonball up and then the cannonball falls down for four minutes of freefall," he once said. Orbit and space "are different leagues." In 2007, he had whipped out a notepad to calculate the difference for an interviewer. And now, on Twitter, he was again playing the role of Professor Musk, delivering a physics lesson:

"It is, however, important to clear up the difference between 'space' and 'orbit,'" he wrote. "Getting to space needs ~Mach 3, but GTO (geostationary transfer orbit) requires ~Mach 30. The energy needed is the square, i.e. 9 units for space and 900 for orbit."

To get to orbit required a massive amount of energy so that the outward acceleration of the spacecraft balances out the force of gravity and essentially falls around Earth. Given the massive amount of velocity required to get an object into orbit—the space station flies at 17,500 mph and circles the globe every 90 minutes—it makes it that much more difficult to land an "orbital-class" rocket. As Musk once said, "You need to unwind that energy in a meteoric fireball, and if there's one violation of integrity, you're toast."

Musk's tweets caused a frenzy in the media, which wanted a response from Bezos to keep the rivalry going, to let these two billionaires in space duke it out. But Bezos kept quiet. The tortoise wasn't going to respond to the hare—at least not yet.

TWENTY-EIGHT DAYS AFTER New Shepard's landing, Musk jumped outside the launch control center onto a causeway at Cape Canaveral, and set his eyes on the launchpad about a mile away. This one he was going to watch live. There was just too much at stake. It was the first launch since the Falcon 9 blew up, and the first since his Twitter taunting.

The company could survive one failure; two would be devastating. Musk was also anxious because he was going to make another landing attempt—this time on land—a chance for him to deliver on his promise that he'd be able to pull off one of his own.

In the days leading up to the return to flight, on December 21, 2015, things weren't looking good for a launch, let alone such an audacious landing. SpaceX was forced to delay by a series of technical glitches related to the temperature of its liquid oxygen fuel, which the company was struggling to keep at an unusually low temperature of minus 350 degrees Fahrenheit.

The low temperature was part of an innovative new version designed to give the rocket a higher performance. It was supercooling the fuel to make it more dense. The denser the fuel, the more SpaceX could pack into the rocket. The more fuel, the more power it could generate.

Another landing attempt would need every ounce of fuel that could be jammed into the booster to enable the engines to fire again on the return. But keeping the fuel at such a low temperature was something new for the company, and could pose problems.

Then there was a glitch with a valve, requiring an adjustment of the ignition sequence by 0.6 second. This was a new, upgraded rocket SpaceX was trying out for the first time since the explosion—more powerful, yes, but also immature, a young buck of a rocket that apparently was fidgeting at the gate.

It was getting closer to Christmas, and some in the industry predicted Musk would have to delay until after the holidays. But he was under pressure from his customer, a commercial communications company, to launch its eleven satellites into orbit by the end of the year. Despite the delays, he remained confident SpaceX would be able to pull it off.

So, here he was at eight thirty on a cloudy, drizzly night on the Florida Space Coast, listening to the flight commander count down. Then came the roar of the engines, the fire, the plume of smoke, and finally, "We have lift off of the Falcon 9," said the announcer on SpaceX's live web broadcast.

About a mile from the launchpad, SpaceX had built a first for Cape Canaveral—a pad that resembled a massive helicopter landing zone, with the X from the company's logo marking the target where the rocket should land. The site happened to be located right next to the launch site where John Glenn had become the first American to launch into orbit during the Mercury program. It was a sacrosanct stage for a potentially historic space feat that would solidify SpaceX's status as the darling of the commercial space flight industry, and Musk as its pied piper, leading his merry band of rocketeers past a threshold no one thought was possible.

Even though he projected confidence in his Twitter lashing of Bezos, Musk would later say he was only 60 to 70 percent sure he'd be able to pull off such a difficult landing. The choreography for this particular maneuver was daunting and complex.

After powering the rocket into orbit, the first-stage engines would cut off after just 2 minutes and 20 seconds. The first and second stages would separate four seconds later, while traveling some 3,700 mph, 50 miles above Earth. Then, as the second-stage engine ignited on its way to orbit, the first stage would fire its nitrogen thrusters, flipping the booster so that it was now facing the opposite direction—flying tail end first.

Then it would reignite three of its nine engines in what was known as the boost-back burn, acting like a giant brake and leaving skid marks in the sky, until the booster started flying back in the opposite direction toward the Cape.

The preprogrammed GPS coordinates in its computers would aim it at the landing site. As it would fall through the increasingly dense air, it would deploy what are called grid fins, wafflelike small wings just 4 by 5 feet wide, which would be used to manipulate the air like a kid putting his hand out of a car on the highway.

Then, the booster would just fall, plummeting like a perfectly poised diver, piercing the clouds as the ground rose. The rocket would then fire its engine once again in what was called the landing burn, as the GPS system would orient the craft toward the landing zone.

SpaceX compared it to "trying to balance a rubber broomstick on your hand in the middle of a wind storm."

No wonder people said it was impossible.

The Federal Aviation Administration had signed off on the landing, granting SpaceX a license to go for it. The air force had, too, but its controllers were on station just in case. If there were any sign that the rocket was going off course, such as about to head toward downtown Titusville, they could blow the thing up remotely, letting the pieces fall into the Atlantic.

Even so, the Brevard County Emergency Operations Center upped its alert state to Level 2—its second highest—just in case. Adding to the drama was the fact that SpaceX was live broadcasting the launch and landing on its website—reality TV of another kind, where the thousands who tuned in would witness either triumph or failure. It was a huge risk. Failure would broadcast a giant fireball that would surely be played, and replayed, by the media.

It stood in stark contrast to the controlled way that Bezos had announced the landing of the New Shepard booster, which had happened the day before the company's PR messengers woke up journalists to deliver news that by then was nearly twenty-four hours old—and now packaged in a scripted press release and a slick, produced video.

The tortoise might have been deliberate and careful. But the hare was letting it hang out for everyone to see, writing the script live and in public, not knowing the outcome. The hare might have been brash and, at times, offputting. But it had guts.

AFTER THE FALCON 9 made it safely to space, Musk stayed out on the causeway, waiting for it to reappear. About ten minutes later, it did. At first, a distant glimmer, small like a streetlight illuminating the fog on a cloudy night. It lowered as if on a string, as SpaceX employees on the Cape gasped and broke into tears. Those gathered at the company's headquarters in Hawthorne, California, just outside of Los Angeles, cheered wildly.

"History in the making," one of SpaceX's commentators said on the livestream broadcast.

Musk watched from the causeway, and could hear and feel what the others at the headquarters could not. An ugly, massive boom that thundered with the force of an explosion hit him like a punch in the chest. Musk assumed the worst.

"Well, at least we got close," he said to himself.

STANDING ON THE causeway, Musk waited for the fireball that surely would follow the boom that had made him think the rocket crashed.

It never came.

He rushed inside the launch control center, where people were cheering what they could clearly see on their computer screens: the Falcon standing triumphantly on the launchpad. "The Falcon 9 has landed," the launch conductor announced. The concussive blast Musk had heard had been a sonic boom, not an explosion.

Whether intentional or not, the words used by SpaceX—"The Falcon 9 has landed"—echoed Neil Armstrong's "The Eagle has landed," after the spacecraft touched down on the moon.

At the company's headquarters in California, it was pandemonium, with hundreds of employees hugging one another, and jumping up and down as if they had just won the Super Bowl. It was like the unbridled celebration at Blue Origin after its landing—only bigger, with more employees. Sitting in the front row of the mission control center, company president Gwynne Shotwell had her hands above her head, touchdown style, and hugged everyone around her. The throng of employees outside the glass-encased control room broke into spontaneous chants of "USA! USA!"

It was perhaps an odd choice for a cheer—this was the feat of a single, private company, not a nation. But in the unbridled exuberance of the moment, it also felt right, as if what they had accomplished extended far beyond the company's headquarters. It reflected an optimism for the future, an affirmation that the impossible goal they had long been working toward might not exist solely in the outsize imagination of one billionaire's head. And it channeled the enthusiasm and similarly grand ambitions of another generation, forty years earlier, that had pulled off what many said could not be done.

For Musk, it was the validation that what he had been talking about for years now might actually be possible: "It really quite dramatically improves my confidence that a city on Mars is possible," he said. "That's what this is all about."

The partying continued until dawn. Musk showed up at the Cocoa Beach site by the beach after visiting the landing pad, still wearing his reflective vest and hard hat. He was greeted like a hero by drunken employees, many in their twenties and thirties, giving out high fives and hugs, soaking it all in, a smile plastered on his face.

In Hawthorne, they partied hard as well. Shotwell dubbed herself the "party mommy" to make sure everyone got home okay. "I was trying to be the grown up managing all the people celebrating the launch," she recalled. "That was a challenge."

Nothing could diminish this high. But underneath the joy, they were furious that their new rival—Bezos—couldn't help taunting Musk, the way Musk had taunted him.

Blue Origin had always practiced an extreme form of discipline, enforced by its ubiquitous nondisclosure agreements and Corleone-like "it's strictly business" ethos.

But now it was personal, if not for Musk's "not quite rarest" tweet, then for the "unicorns dancing in the flame duct" crack. Or any other of the indignities and perceived slights that had fueled what was now a full-blown rivalry.

"Congrats @SpaceX on landing Falcon's suborbital booster stage," Bezos tweeted soon after the landing. "Welcome to the club!"

Whether it was meant sincerely or not, it came off as a counterpunch: he had done it first.

As the tweet spread, SpaceX employees were increasingly angry, as was Musk.

"That was a pretty snarky thing for him to say," Musk said later.

Shotwell said she "rolled [her] eyes and kept quiet. It was a silly thing for him to say."

But before Musk could go on a rampage, his team showed him what was happening on Twitter: their fans were retaliating for him. They had understood from Musk's repeated statements that space was not orbit, and they responded to the slight with a call to arms.

"@JeffBezos @SpaceX not even in the same league buddy. Nice try."

"@JeffBezos @SpaceX If you want to gain supporters—be gracious. Perhaps being an antagonistic dick is not the way to go?"

"@JeffBezos @SpaceX enough said," one tweeted with an image of the companies' rockets, side by side, designed to illustrate how the endowed Falcon 9 made the New Shepard look prepubescent by comparison.

Once Musk saw the reaction on Twitter, he recalled, he relaxed and decided that "I'm not going to respond to such absurdity," especially after the "Internet spanked him pretty hard for that one."

It was all good. There was a rocket standing tall on the landing pad. There would be no tweet storm tonight.

BRANSON HAD WATCHED the budding rivalry between his fellow billionaires unfold from a distance, and cringed. He was all for impulsivity, and living in the moment, but he made a point of avoiding conflict. This was distasteful, a war that no one could win.

"Rivalry is generally good, I mean definitely good from a consumer point of view," he said. Over the years, he and Musk had formed a bond—"I'm friends with Elon, and know him relatively well. He comes down to Necker on occasion." But he wouldn't meet Bezos until years later, and when asked about the Twitter spat, he paused, searching for the right tone.

"Tweets are not necessarily becoming if you use them for rivalry purposes," he said, then catching himself. "Anyway, I'd rather stay above the fray."

The best way to compete was to stay focused on your own product, not your competitor. And with the breakup of SpaceShipTwo, Branson's had taken a blow. But now, more than a year later, he was back. His team had rebuilt SpaceShipTwo, making it safer and more reliable, and he was again ready to show it off.

HE MADE HIS entrance standing through the sunroof of a snow-white Land Rover, blowing kisses and waving to the crowd like a triumphant Caesar riding a chariot into the Colosseum. The last time Branson had appeared publicly in Mojave, his spacecraft had shattered into pieces on the desert floor. Now, he had a new spaceship to christen in a baptism that he hoped would cleanse away the pains of the past failure and reinvigorate new hope.

The National Transportation Safety Board had wrapped up its nine-month investigation, concluding that a "lack of consideration for human factors" led to the midair breakup of the spacecraft. It found that Scaled Composites, which had built the vehicle for Virgin Galactic, had failed to properly train its pilots and did not implement basic safeguards to prevent the human error that caused the accident.

Yes, Michael Alsbury, the pilot killed in the crash, had unlocked the feather system prematurely. But he never should have been able to do so, the safety board said, and the company's failure to even consider that possibility was one of a series of systematic failures that had led to the crash. As board member Robert Sumwalt said, Scaled Composites "put all their eggs in the basket of the pilot doing it correctly." Unfortunately, humans inevitably make mistakes, "and the mistake is often times a symptom of a flawed system."

Virgin Galactic responded by implementing an inhibitor that would prevent the pilot from unlocking the feather prematurely. And it fired Scaled Composites, saying it would build the spacecraft itself.

"From now on, we've taken everything in house and anything that happens from today will be down to Virgin Galactic," Branson said.

Now, on February 19, 2016, he had a new SpaceShipTwo to unveil. And his entrance—to the requisite pumping music, swirling lights, and chilled champagne—fulfilled the Branson stereotype and satisfied those who had come expecting a red-carpet show from one of the world's most celebrated playboys. Branson was more than happy to play the part, with his rock star leather jacket and jeans, flowing hair, perma-white smile, and British charm. Harrison Ford, Han Solo himself, was sitting in the front row. But the real star power of the day came from the world's premier celebrity physicist, Stephen Hawking.

Inside the company, its officials were worried about coming off as too cavalier in the wake of the fatal accident. They wanted to put on a show glamorous enough to erase some of the pain, and to restore confidence, but they were also very cognizant of crossing a line—especially when they were years behind schedule and had still not flown a single paying customer. The crash and the subsequent delay also meant that Spaceport America, the futuristic facility that had cost taxpayers $220 million to build, continued to languish in the New Mexico desert, waiting for Virgin Galactic to fly.

The company's mishap had consequences far greater than an empty spaceport that had been a drain on government coffers; it had cost a life, and now Virgin's executives had to show they were sober and serious in their pursuits. Hawking, then, was a brilliant choice, a sign of restraint. Although he was unable to travel to the event because of illness, his distinctive, computerized voice filled the hangar.

"I have always dreamed of spaceflight," Hawking said. "But for so many years I thought it was just that—a dream. Confined to Earth and a wheelchair, how could I experience the majesty of space except through imagination and my work in theoretical physics?"

He said that, years before, Branson had offered to give him a ride to space, and added, "I would be very proud to fly in this spaceship."

Instead of trying to erase the past, Virgin Galactic embraced it. One executive choked up when discussing Alsbury's death and legacy. And the company's chief executive, George Whitesides, didn't shy away from it, either.

"It's now been sixteen months since our flight test accident. That was a hard day," he began.

He recalled meeting Branson in "this very hangar" just after the crash. "It was a moment when years of hard work were put into public doubt, and the life of a brave test pilot, a family man, and a friend to many of us, was lost. We walked through this hangar, and we stood in front of the partially built serial number two space ship. It was positioned right over there. And we wondered, does this collection of carefully constructed parts represent our past—or our future?"

The answer, in this choreographed event that trod the fine line between celebration and memorial, between rebirth and the funereal, was obvious. But as the company charged forward, it was clear that the once immature and impatient rush to the cosmos had been slowed down and chastened, with a renewed sense of care replacing exigency.

Even before the event, in an effort to manage expectations and assure potential customers that it was moving deliberately and making safety paramount, Virgin Galactic released a statement warning: "If you are expecting SpaceShipTwo to blast off and head straight to space on the day we unveil her, let us disillusion you now: this will be a ground-based celebration."

For a company built on exploding expectations rather than managing them, it was an extraordinary statement. Branson and his Virgin brand had never been in the business of "disillusionment," but of making the illusory real. Death, however, was sobering, and Virgin Galactic was faced with the delicate balance of promoting its newest spacecraft, and the once unthinkable prospect of routine space travel, against the dangers and difficulties inherent in that endeavor.

First, the new spacecraft would go through a series of rigorous tests, and even before the vehicle was assembled, the company would lay out how it "poked, prodded, stretched, squeezed, bent, and twisted everything to be used to build these vehicles." It was as if Virgin Galactic was unveiling a baby's car seat, not a spacecraft.

People who know Branson well often said the playboy image was something of a myth. That he is, at heart, a family man, surprisingly earnest and disarmingly self-deprecating. Unlike Musk and Bezos, who liked nothing more than to rattle on about the technical aspects of their rockets, Branson would come across as a bit insecure, making sure engineers were close by to answer any detailed questions. He provided the vision, not the technical specifications.

Branson may have still been a playboy, but he was also a grandfather now, sixty-five years old, and at the unveiling he was surrounded by four generations of his family. His mother, nearly one hundred, was in the audience, as were his son and granddaughter, who was celebrating her first birthday.

In the back of the hangar there was plenty of champagne, but Space-ShipTwo would not be christened with bubbly. Instead, the Bransons huddled around the newest member of their family, Eva-Deia, an innocent cherub with bright eyes and blond hair, and baptized the spaceship with the baby's milk bottle.

IN THE MONTHS that followed, Musk and Bezos started to play nice, at least in public. Their Twitter spat had touched off an irresistible media frenzy that pitted the pair against each other—a pair of tech billionaires fighting for cosmic domination—a made-for-large-font headline neither wanted.

For someone who cultivated his image as meticulously as Bezos, it wasn't dignified even to be perceived as feuding with Musk. When competitors came after Amazon, it only drove him to want to succeed even more. That was as true in the world of online retail as it was in space. Bezos would take the high road and remain focused on the immense challenge of getting off the planet, just as at Amazon he urged his team to stay relentlessly focused on the customer.

"At Blue Origin, our biggest opponent is gravity," he said during an awards ceremony. "The physics of this problem are challenging enough. Gravity is not watching us and saying, 'Uh-oh those Blue Origin guys are getting really good. I'm going to have to increase my gravitational constant.' Gravity doesn't care about us at all."

The cosmos stretched far and wide, with plenty of room for lots of companies to live long and prosper. The business of space didn't have to be a zero-sum game.

"Oftentimes, it's very natural to think of business competition like a sporting event," Bezos said during a Q & A with Alan Boyle of Geekwire at an annual space conference in 2016. "Somebody leaves the arena a winner, and somebody leaves the arena a loser. In business, it's usually a little different from that. Great industries are usually built by not just one or two or three companies, but usually by dozens of companies. There can be many winners, even hundreds and thousands of companies in a truly great industry. I think that's where we are headed toward here.

"From my point of view, the more the merrier. I want Virgin Galactic to succeed. I want SpaceX to succeed. I want United Launch Alliance to succeed. I want Arianespace to succeed. And, of course, I want Blue Origin to succeed. And I think they all can."

While it bothered Musk when people compared the accomplishments of SpaceX to Blue Origin's, he, too, became more conciliatory.

"In general, I think it's important that we advance spaceflight for the good of humanity," he said. "If I could press a button and make Blue Origin disappear, I would not press that button. I think it's good Jeff is doing what he's doing."

They were driven by the business opportunities in space, by adventure, and by ego—imagine the Promethean legacies they'd leave after opening up the Final Frontier.

But there was no motivator quite like head-to-head competition. No one knew this better than Musk and Bezos. Amazon wouldn't have become Amazon if it didn't have Barnes and Noble to set its sights on. Tesla wouldn't have been Tesla if it wasn't taking on all of Detroit. And SpaceX, from its inception, had targeted the Alliance, seeking to disrupt the cushy monopoly it had held for years and crowbar open the lock it had on the Pentagon's golden chest.

Competition had driven the original space race. Without the Soviets threatening to own the ultimate high ground, the United States would have never made it to the moon. After the Soviets had made Yuri Gagarin the first man to orbit Earth, President Kennedy had been anguished, running his hands through his hair and nervously tapping his teeth with his fingernails during a meeting at the White House.

"If somebody can just tell me how to catch up. Let's find somebody—anybody. I don't care if it's the janitor over there," Kennedy had pleaded, adding later that "there's nothing more important."

Less than a decade later, as Neil Armstrong crossed the finish line, the first man to walk on the moon was magnanimous, proclaiming the victory as "one giant leap for mankind."

The race complete, the victor triumphant, the loser vanquished, there was then a long fade in human spaceflight, a retreat even. The lack of competition led to complacency. A comfortable wither. Despite the repeated promises of presidents hoping to channel Kennedy and summon a "because they are hard" call to arms, the next giant leap—Mars, moon bases, a civilization in the stars—never came. Hope and dreams may have sounded great at the podium. On the launchpad, they only went so far.

If Musk and Bezos were going to be the true heirs to Apollo, if, at long last, they were going to push humans further into the cosmos, building that railroad system to the stars, they would have to crouch down alongside each other, get on their mark, get ready, and go. One eye focused clearly on that distant, impossible goal; the other, on the competitor just over their shoulder.

For all the conciliatory talk, the truth was they needed each other.

Rivalry, it turned out, was the best rocket fuel.

14

Mars

THE FAITHFUL STARTED lining up hours before the show was to start. The first ones huddled early at the conference hall in Guadalajara, Mexico, in small clusters as die-hard adherents to the vision. They were like groupies camped out for tickets to the new Star Wars movie, making the wait part of the May-the-Force-Be-with-You ritual, a curbside tent party with costumed storm troopers, Han Solos and Yodas.

Behind the locked doors, their Yoda was getting ready. Elon Musk wanted to get the details right. This was his big moment, and he wasn't going to rush it. For months, he'd been teasing the speech he was to give today, September 27, 2016, at the International Astronautical Congress, an annual space conference. The hype had built to a fever pitch to the point where Musk had overshadowed everything else, turning the days-long international space convention into the Elon Show.

When he founded SpaceX a lifetime ago in 2002, Musk was an unknown eccentric with a wild idea to privatize space that even he didn't think would work. Now, he was a worldwide celebrity, Tony Stark in a Tesla, with a space company that had attracted a cultlike following that clicked on the company's launch and landing YouTube videos by the millions.

SpaceX had transcended corporate America the way NASA had once transcended government bureaucracy, becoming an institution of hope and inspiration. Now Elon—always the one name—was the new face of the

American space program, the embodiment of exploration, a modern-day amalgam of JFK and Neil Armstrong, with 10 million Twitter followers.

The press room in Guadalajara was overflowing with reporters who had come from all over the world for this long-awaited speech, titled "Making Humans a Multiplanetary Species," in which Musk would, finally, lay out his plan to colonize Mars.

In the months leading up to Guadalajara, he disclosed some of the details, telling the *Washington Post* that he intended to build a transportation system to the Red Planet like the railroads that traversed the United States, with the goal of the first humans landing on Mars in 2025. NASA had already announced that it would partner with SpaceX to fly its Dragon spacecraft, without any passengers, to the surface of Mars. Then, every two years after that, when the orbits of Earth and Mars were at their closest points, SpaceX would send additional supplies, all ahead of what would become the first human settlement.

"Essentially what we're saying is we're establishing a cargo route to Mars," he told the *Post*. "It's a regular cargo route. You can count on it. It's going to happen every 26 months. Like a train leaving the station."

Mars had been the goal since SpaceX's inception, the reason he founded the company. Steve Davis, one of SpaceX's earliest employees, remembered memos he received from Musk as early as 2004 that asked, "How much propellant do we need to land on Mars?" And at his first performance evaluation review, they didn't talk about Davis's performance: "We talked about Mars. That was the entire conversation. How do you get to Mars?"

Now it was coming together, at least in Musk's mind. In the interview with the *Post*, Musk was so excited about the prospect that he could barely contain himself. "I'm so tempted to talk more about the details of it," he said, catching himself, not wanting to scoop his Guadalajara speech. "But I have to restrain myself."

For the first missions, SpaceX would launch its Falcon Heavy rocket, a twenty-seven-engine beast that was essentially three Falcon 9s strapped together. But for the human colony, it would build the Mars Colonial Transporter, or what was known inside SpaceX as the BFR—Big Fucking Rocket.

"BFR is a pretty good name for this, it's very big," he told the *Post*. "Obviously I want to reserve the details of that for September. But this is going to be mind blowing,"

"Mind. Blowing," he repeated, in rhapsodic tones. "It's going to be really great."

Now the moment had finally arrived. Finally, the doors to the convention hall opened and the crush ran frenetically for seats close to the stage. For a few minutes, it was so chaotic that those with seats filmed the people swarming around them.

Finally, it calmed down and Musk, wearing a dark blazer and white shirt and with a several-day-old beard, took the stage, standing before a giant, illuminated photo of the red planet.

"So," he said, "how do we figure out how to take you to Mars?"

IN THE DAYS before the speech, many thought Musk would cancel. That even he didn't have the temerity to go through with it. Not when, just three weeks before, another of his rockets had blown up in a spectacular fireball—and SpaceX still didn't know why.

This time, the explosion happened as the rocket was being fueled on the launchpad in preparation for an engine test fire, days before it was to launch. But something went horribly wrong and suddenly the rocket detonated, sending a plume of thick, dark smoke billowing over the Florida coast. No one was injured, but the blast could be felt for miles.

For the second time in just over a year, SpaceX had lost a rocket in a disastrous explosion. But this time, instead of carrying cargo to the International Space Station, its payload was a $195 million Israeli satellite that was going to be used, in part, by Facebook to beam the Internet to developing nations, particularly in sub-Saharan Africa.

The explosion was bad enough. But in an effort to save time, SpaceX had loaded the multimillion-dollar satellite on the rocket ahead of the test fire, a decision that in retrospect looked foolish and needlessly risky. Standard practice, and perhaps common sense, said to wait until after the fiery test, when the rocket had been cleared to go. But SpaceX liked to push the envelope, and was trying to move fast to get through its backlog of launches.

Mark Zuckerberg, Facebook's chief executive, was in Africa when he got the news. "As I'm here in Africa, I'm deeply disappointed to hear that SpaceX's launch failure destroyed our satellite that would have provided connectivity to so many entrepreneurs and everyone else across the continent," he wrote on Facebook.

On the *Late Show* the following night, Stephen Colbert showed a clip of the explosion. "BOOOM! It blowed up real good!" he said. "Now, it was unmanned, okay. Very important: No one was harmed. But it was carrying a satellite that was supposed to bring the Internet to sub-Saharan Africa. So, yeah, I know now not only do they not have clean drinking water, but they can't bitch about it on Yelp."

As investigators began their probe into the cause, SpaceX was grounded again. After the explosion the year before, Musk had been confident shortly afterward that the company knew what had gone wrong and how to fix it. But this was a mystery. What causes a rocket suddenly to blow up while just sitting on the pad?

Musk was apparently dumbfounded, and a week after the explosion vented on Twitter that the loss of the rocket was "turning out to be the most difficult and complex failure we have ever had in 14 years."

He added that it was "important to note that this happened during a routine filling operation. Engines were not on and there was no apparent heat source." He pleaded with the public to come forward with any recordings of the explosion. He then added an element of intrigue by saying that the investigators were "particularly trying to understand the quieter bang sound a few seconds before the fireball goes off. May come from rocket or something else."

On the Internet, where conspiracy theories were already percolating, some speculated that the "something else" was a projectile, maybe even a bullet or UFO. On Twitter, Musk was asked about the possibility of something hitting the rocket, and he fueled speculation even further by saying, "We have not ruled that out."

Although they didn't say so publicly at the time, SpaceX investigators were looking seriously at sabotage.

"We literally thought someone had shot the rocket," Musk recalled later. "We found things that looked like bullet holes, and we calculated that someone with a high-powered rifle, if they had shot the rocket in the right location, that exact same thing would have happened."

At first, the company was baffled about the cause of the explosion and so, "the first thing you do is think it's some outside force, right," said SpaceX president Gwynne Shotwell afterward. "Because we couldn't figure out how in the world this could have happened."

If someone did shoot the rocket, SpaceX knew it needed to collect whatever evidence it could as fast as possible. "So, for sure, we put pressure on the air force and the FAA to go collect whatever forensic data was possible," Shotwell said.

Early indications were that something caused an upper-stage helium bottle to explode, but at the SpaceX test site in McGregor, Texas, engineers were trying to replicate the explosion. But "we were having a hard time blowing these bottles up," she said.

So, instead, they got a rifle, "and we shot it," she said. "And the signature on the bottle was just like the signature on the bottle that we recovered. That was an easy test to do. It's Texas, right, everybody's got a gun and you can blow stuff up."

With SpaceX hobbled, the United Launch Alliance pounced, hoping to poach some of SpaceX's customers by offering an expedited launch schedule to get its payloads in orbit. And it boasted an impressive track record, one its new competitor could not match: more than one hundred consecutive launches without a single failure.

"The priorities of all of our customers include ensuring their spacecraft launches on schedule, securing the soonest possible manifest date and completing the mission with 100 percent success," said the Alliance's chief executive and president, Tory Bruno, in a statement. "To address these priorities, we have been working on this offering for more than a year, which allows our customers to launch in as few as three months from placing their order."

Competition between the companies was as intense as ever. After years of lawsuits, SpaceX had finally won the right to bid on the Pentagon's lucrative national security launch contracts. For a decade, the Alliance held a monopoly on that work, but now SpaceX was threatening the Alliance's main revenue stream.

But two weeks after the Falcon 9 rocket exploded, the long-running feud between them took a bizarre twist. A SpaceX employee suddenly showed up at one of the Alliance's facilities on Cape Canaveral with an odd request: Could he have access to the roof?

The reason, he explained, was that SpaceX had still images from a video that appeared to show a shadow, then a bright white spot, coming from the roof. The Alliance's building was about a mile away from the launchpad and had a clear line of sight to it.

Although the SpaceX employee was cordial and not accusatory, the implication of sabotage was explosive, and to the Alliance, incredulous. The latter refused to let the SpaceX employee in, and called the air force, which found nothing amiss.

As Musk took the stage in Guadalajara, the mystery continued, as did the conspiracy theories.

"So, how do we figure out how to take you to Mars?" Musk asked during the introduction to his speech.

The answer was: a big rocket. To demonstrate just how big, Musk's presentation started with a drawing of a person standing next to it. As the view zoomed out to capture the size of the rocket, the person got smaller and smaller in comparison until he was barely perceptible.

"It's quite big," Musk deadpanned.

Since it would be able to go beyond Mars, some were now calling what had been known as the Mars Colonial Transporter, the Interplanetary Transport System.

Whatever its name, it was a behemoth, more than 400 feet tall, with forty-two engines and a spacecraft that could carry one hundred people or more, refuel in orbit, and then cruise to Mars at 62,634 mph before touching down. On stage, Musk painted an optimistic portrait of the future, vowing that a "self-sustaining city on Mars" of a million people could be achieved within forty to one hundred years of the first flight, and he talked about a "Mars Colonial Fleet" departing en masse for the Red Planet.

The trip there has "got to be really fun and exciting," he said. "It can't feel cramped or boring. But the crew compartment, or the occupant compartment, is set up so that you can do Zero-G games, float around. It would be like movies, lecture halls, cabins, a restaurant. It will be, like, really fun to go. You are going to have a great time."

The highlight of his presentation was a fantastic four-minute video that showed the massive booster taking off, and the spacecraft cruising through space until Mars showed up in its massive windshield as an inviting red and golden orb, a promised land with an atmosphere glowing around it like a halo.

The video ended with a time lapse of a desiccated, dead planet, which scientists believed could have once supported life, transforming into an Earth-like planet, with green and blue oceans replacing the vast red deserts of Mars.

Musk was right. It was mind-blowing. All of it. The supersize rocket. The spacecraft with a windshield that showed the Red Planet looming large. The idea that Mars could be heated up and made habitable, like Earth.

But it was also all a bit too fantastic, crossing into the surreal with the same characteristics as good science fiction: "plausible but not probable," as one noted space expert later observed. Especially since, at that very moment, SpaceX's launchpad was in ashes and its rocket was grounded. Not to mention that the company had only flown satellites and cargo and had yet to fly a single person anywhere, not even in low Earth orbit, let alone Mars.

The timeline was almost laughably improbable—with first flights in 2018 on the Falcon Heavy, a rocket that had endured repeated delays and technical problems and had yet to fly. Mars remained an exceedingly difficult mission. On average, it's 140 million miles from Earth, though the two planets come much closer to each other every twenty-six months. And of the forty-three robotic missions to Mars, including flybys, which were attempted by four different countries, not companies, only eighteen were deemed a complete success.

If Musk were going to be able to pull this off, it "would be a gigantic human engineering endeavor, greater in scope, scale, and cost than the Manhattan Project," said Gentry Lee, the chief engineer for solar system exploration at NASA's Jet Propulsion Laboratory.

For a human Mars colony on Musk's timeline to be successful, it "would have to develop and infuse new technologies at a much faster rate than we have ever achieved before on any project."

The experts weren't the only ones who were skeptical of Musk's dream. Loren Grush, a top space journalist working at the Verge, pressed Musk after the presentation on a key detail that he hadn't addressed.

"You didn't touch much on how you will keep humans safe on the way over there for either deep space radiation or how they will live on the planet," she said. "Can you give us some insight into the life support system, habitats, stuff like that?"

For someone as detail-oriented as Musk, he responded with an almost blasé approach, ignoring the crux of the question and simply saying that "the radiation thing is not too big of a deal."

Another big question hanging over the presentation was: Who was going to pay for all of this? Musk said he would "make the biggest contribution I can" of his own personal wealth. But at one point he joked that SpaceX might have to use Kickstarter, the online fund-raising platform, to raise money.

"As we show this dream is real . . . I think the support will snowball over time," he said, without offering any details.

But that was more of a wish than a concrete business plan. And his idea that this would ultimately have to be a "public-private partnership" also seemed improbable. NASA had its own plan to get to Mars, and it was building its own rocket and spacecraft to get there.

Although Obama killed the Bush-era Constellation program, Lockheed Martin's Orion crew capsule had survived the chopping block. In 2014, it flew to an altitude of 3,600 miles, farther than any spacecraft designed for humans had traveled in more than forty years. Although no astronauts were aboard for its maiden test flight, NASA cheered the mission as a "new era" in human space exploration.

But soon afterward, government watchdogs were warning that the same sorts of cost overruns and schedule delays that had plagued the Constellation program were weighing down the Obama administration's plans as well. The new rocket that replaced the Bush-era Ares V was known as the Space Launch System (SLS). But it had yet to fly and was derided by critics as the "Senate Launch System" because it appeared like it was more of a program designed to create jobs in congressional members' districts than actually fly to Mars.

The $23 billion SLS/Orion program came as NASA faced "struggles with poor cost estimation, weak oversight and risk underestimation," the US Government Accountability Office warned. NASA's Mars mission, slated for sometime in the 2030s, seemed so implausible that veteran space journalists, weary of NASA incessantly promoting its "Journey to

Mars," openly mocked it as a journey to nowhere. Support in some corners of Congress was waning.

"We made a wrong decision when we went down this road," said Representative Dana Rohrabacher, a California Republican.

But that didn't mean they were ready to just cancel NASA's SLS/Orion program and fund Musk's Mars mission instead, even if he was promising to get there first.

FOR WEEKS, MUSK and a small SWAT team of SpaceX employees had spent their Saturdays working on the Mars architecture, and the presentation. But they seemed to overlook a key detail—the Q & A session that would follow.

The organizers left the microphones accessible to anyone in the audience, and it soon took a turn into the absurd, as general audience members took the opportunity to ask Musk whatever they wanted. Someone named Aldo, who sounded as if he was stoned, said he had just gotten back from Burning Man, the annual pilgrimage in the Nevada desert, where he said it was cold, dusty, and uncomfortable—and compounded by overflowing sewage.

"Is this what Mars is going to be like—just a dusty, waterless shitstorm?" Aldo asked.

Another guy told Musk he wanted to give him a comic book about the "first man on Mars, just like you." But he couldn't get by the guards protecting the stage, and asked, "Should I just throw this onto the stage?"

Then there was the woman who asked, "On behalf of all the ladies, can I go upstairs and give you a kiss, a good luck kiss?"

Musk shifted stiffly as the crowd started whooping and hollering, as if at a brothel.

"Thank you," he offered, awkwardly. "Appreciate the thought."

IF THEY DIDN'T take him seriously, many were, at least, inspired. The long lines were a testament to that. As was the rush into the hall, a reignited enthusiasm for space and science and exploration, topics that hadn't galvanized the American public quite like this in a long, long time.

Maybe Musk's Mars mission was an illusion, pure fiction fit for the comic book that the joker in the convention hall tried to throw on stage. But maybe that didn't matter. What if the point of the whole exercise was to make it *seem* real, as if it *could* be done?

SpaceX was, after all, a place where the motto was, as Shotwell had said, "set audacious, nearly impossible goals and don't get dissuaded. Head down, plow through the line—that's very SpaceX."

If achieving the impossible was the goal, then a single rocket explosion was a mere speed bump, not a roadblock—a temporary inconvenience, not a catastrophe.

If the tortoise was content with *slow is smooth and smooth is fast*, the hare was ready to embrace all the virtues of impatience. It had been long enough. The country had wasted enough money trying to repeat Apollo, only to fail time and time again.

Those who knew Musk and the company he had built understood that this moment, the nadir—a launchpad in ashes, rockets grounded, investigators probing, skeptics tsk-tsking, competitors pouncing—was precisely the time to unveil the most audacious plan of all.

Cancel the speech? Never.

SPACEX HAD RECENTLY designed retro-style travel posters showing Jetsons-like tourists on Mars, meandering through Valles Marineris with a jet pack, taking a tram to the top of Olympus Mons, "the solar system's highest peak," and gazing out on the expanse where they could "take a space age cruise of the moons of Mars." They were a mix of marketing and fantasy that went viral, a sign of the emergence of a new leader in human space travel.

For all its accomplishments, NASA could no longer lay exclusive claim to the title. The space shuttle program had been a compromise that didn't deliver on its goal of providing reliable, low-cost access to space. It was pricey and dangerous, taking the lives of fourteen astronauts. Bush's Constellation program, which was supposed to take humans back to the moon, had been killed. NASA's replacement program, which was supposed to get to Mars, didn't seem within decades of doing so with the overbudget, behind-schedule SLS rocket and Orion spacecraft.

Musk, then, filled a void that was larger than his space company.

"His job is to provide inspirational leadership not just for SpaceX, but for the larger space community," said John Logsdon, the noted space historian who had written books about Kennedy's and Nixon's space programs. "There hasn't been someone like that for a very long time."

Or ever. The fact that SpaceX existed at all was an improbable triumph of an aggressive and relentless business strategy, innovative engineering, and, perhaps above all, imagination. The idea that a private individual could start a space company and be successful seemed about as outrageous as what Musk was now trying to accomplish.

As if to prove the point, Musk had showed the audience in Guadalajara a ridiculous picture of the early days of SpaceX, when it had just a few employees and had invited a mariachi band to a company party.

"Just to show you where we started off in 2002, SpaceX basically consisted of carpet and a mariachi band. That was it. That's all of SpaceX in 2002," he said. "I thought we had maybe a ten percent chance of doing anything, of even getting a rocket to orbit, let alone getting beyond that and taking Mars seriously. But I came to the conclusion if there wasn't some new entrant into the space arena with a strong ideological motivation, then it didn't seem like we were on a trajectory to ever be a spacefaring civilization and be out there among the stars."

Now, he had more than five thousand employees working in Hawthorne, McGregor, and Cape Canaveral. The company also had a launchpad at Vandenberg Air Force Base in California and was building its own private launchpad in Brownsville, Texas, where it would be free to launch as it saw fit without worrying about the sometimes hectic schedule of the government-owned facilities.

After SpaceX sued the US Air Force over the right to compete for the national security launches, the parties finally settled and the Pentagon finally certified the Falcon 9 for the missions. Musk had been pursuing the lucrative launches for years, and now it was finally able to compete against the Alliance, which had held a monopoly on the billions of dollars' worth of contracts for a decade.

The victory was the end of what Tim Hughes, SpaceX's general counsel, called a "decade-long grind."

"Our mantra from day one was: allow us to compete. And if we lose on the merits so be it," he said. "But you've got to let competition flourish, and you need to let what we perceive to be the home team, play. And by home team we mean this is an all-American rocket. No reliance on Russian engines or other major component parts."

But the first time the air force held a competition for one of the launches, the Alliance balked, saying it was unable to submit a bid. It claimed that the procurement was set up so that the competition was weighted toward who had the lower price—SpaceX—and not who had the most experience and record of past performance, which would have given the Alliance the edge.

Congress had limited the use of a Russian-made engine that the Alliance used in its Atlas V rocket, and those restrictions made it impossible to bid, the company said.

SpaceX and others weren't buying it.

"I thought it was a coward's move. You can quote me," said Shotwell. "They did not want to lose. And they knew they would lose."

FROM THE MOMENT he founded SpaceX, the hare had been blazing a trail, and along the way clearing the path for others. Musk had gotten NASA to trust companies like his. He had taken on the air force in court and won. He had made space cool again. And he had done it first. In his wake, a new commercial space industry was beginning to emerge in earnest.

Investors, long leery of the risky industry, started to wade in. In 2014, the global space economy totaled $330 billion, a 9 percent jump from the previous year and up from $176 billion in 2005, according to the Space Foundation, a nonprofit space advocacy organization. In 2015, Google and Fidelity invested $1 billion in SpaceX, backing another of Musk's ventures: a bold plan to build a constellation of thousands of small satellites that would swarm over Earth, beaming the Internet to remote parts of the world.

SpaceX had gotten so attractive that the company had to actually turn away money, telling Steve Jurvetson, one of Silicon Valley's most successful venture capitalists and an early backer of SpaceX, to hold off.

"There's so much interest, they can't take it all," said Jurvetson.

By mid-2017, after raising $350 million in a new round of funding, SpaceX was valued at $21 billion, "making it one of the most valuable privately held companies in the world," the New York Times reported.

The prospect of reusable rockets dramatically lowering the cost of launch fueled the growth, as did the revolution in small satellites. For decades, satellites had been big, as large as a garbage truck, and expensive, costing hundreds of millions of dollars. But now the technology had changed, and like an iPhone, they had shrunk in size, to the size of a shoebox, costing far less.

Musk wasn't the only entrepreneur looking to cash in on the new satellite technology. OneWeb, a company backed by Richard Branson, also planned to put up a constellation of hundreds of miniature satellites that it said would connect the billions of people without Internet access to the digital economy.

Google executives Larry Page and Eric Schmidt invested in Planetary Resources, which planned to mine asteroids. Filled with precious metals, the asteroids are the "diamonds in the rough of the solar system," Eric Anderson, the company's cofounder, told CNBC.

Asteroids have "rare metals, industrial metals and even fuels," he said, "so we could create gas stations in space that would enable us to travel throughout the solar system just like *Star Trek*."

It sounded like something out of a James Cameron movie. And maybe it will be. The Hollywood director served as an adviser to the company. But it also is the subject of a law signed in 2015 by President Obama that gives US companies the rights to the resources they mine in space. And it has gotten the attention of investment bankers.

"We believe space mining is still a long way from commercial viability, but it has the potential to further ease access to space and facilitate an in-space manufacturing economy," an analyst for Goldman Sachs wrote in a note to investors. "Space mining could be more realistic than perceived . . . a single asteroid the size of a football field could contain $25 billion to $50 billion worth of platinum."

Robert Bigelow, the multimillionaire founder of Budget Suites of America, had developed a space habitat, made of a Kevlar-like material that inflated, like a balloon, once in orbit. Another venture called Made in Space had put the first 3-D printer on the International Space Station, as it worked toward creating manufacturing facilities in space.

The wave of new companies touched off an Apollo-like renaissance in the top aerospace engineering programs. At Purdue University, applications for the undergraduate aerospace engineering program jumped 50 percent.

"The demand for our program has increased dramatically because of companies like Virgin and SpaceX and Blue," said Stephen Heister, a professor at the School of Aeronautics and Astronautics. "We can only absorb so many people. We have to turn away some who are really very qualified. . . . I'm an old guy. I graduated in the early 1980s, and in my entire career, this is the most exciting time."

MUSK WAS THE benefactor of much of that excitement. He was the face of this new industry, its de facto leader. He was the one on stage, taking unfiltered questions from the masses. He was the hare. *Head down. Plow through the line.* Anyone who followed, Blue Origin included, owed a measure of their success to SpaceX and its relentless march toward Mars.

As Musk had told the *Washington Post*, his goal was to reignite people's interest in space, to "get people fired up." Mars, he said, would be "the greatest adventure ever." In Guadalajara, his goal was to "make it seem possible. To make it seem like it's something we can do in our lifetimes. That you can go."

He would take you for the low, low price of $200,000. The ticket price that Branson had originally charged for suborbital jaunts to space would, one day, be enough to get you to Mars—return flight included. Of course, it would be hard and dangerous and, as Musk had said, "people will die."

But like all of the great dreams, as Superman actor Christopher Reeve said, it would first seem impossible, then improbable, and then inevitable. You just had to believe and see through the dense forest of disbelief to a point in the distance where doubt gave way to an improbable question: What if everything Musk was saying was true?

15

"The Great Inversion"

THERE WAS NO sign on the outside. No company logo. Nothing but the address on a nondescript warehouse. Inside, past the front desk attendant, who asked whether you'd been here before and whether your nondisclosure agreement was already on file, and up a flight of stairs, visitors were greeted by a scattering of space memorabilia that was more like an eccentric museum collection than a corporate lobby.

In the center of the hardwood floor was a model of the Starship Enterprise used in the original *Star Trek* motion picture. There was a Russian spacesuit on display, a proposed space station that was never built, and a model of a domed habitat, as if on Mars. There was a poster of a monster rocket engine, and a nod to the past: an anvil ca. 1780 from Troyes, France.

On the walls of Blue Origin's lobby were inspiring quotes, including one from Leonardo da Vinci: "For once you have tasted flight you will walk the Earth with your eyes turned skywards, for there you have been and there you will long to return."

But the centerpiece of Jeff Bezos's collection was a rocket ship model, shaped like a bullet, which stretched up to the open floor above. A Jules Verne–inspired, Victorian-era vision of a rocket with room for five and engines directed to a fire pit below, making it seem as if it were blasting off from the lobby. Inside were plush, velvety couches, a bookshelf with *20,000 Leagues Under the Sea* and *From the Earth to the Moon,* a stocked whiskey cabinet, and a pistol. It was a quirky and detailed accoutrement, designed to make the foreign familiar.

The lobby was a great big Valentine to adventure, the early days of the Space Age and the intersection of science fiction and art, a childhood dream come alive. If a rocket-cum-employee lounge wasn't enough to convince visitors they had entered a strange and curious place—where employees' dogs roamed about freely—there was the corporate coat of arms, as if Blue Origin was laying down its heraldry for future generations, not on a shield but on the wall, like a mural.

It was an involved piece of art, loaded with trippy symbols from Earth to the stars with the velocities needed to reach various altitudes in space. There were a pair of turtles gazing skyward, an homage to the winner of the race between the tortoise and the hare, celebrating the deliberate and methodical approach. But there was also an hourglass symbolizing human mortality, and the need to move expeditiously.

Before he had made his fortune at Amazon, Bezos had been outbid at the Sotheby's auction of space memorabilia. But in the years since, he had more than made up for it. Here was a Mercury-era NASA hard hat, an Apollo 1 training suit, and a heat shield tile from the space shuttle.

Then there was the curious piece of art tucked away in the corner. It was composed of 442 spools of thread stacked vertically on top of one another, as if in a spice rack. Together they looked like nothing more than the inside of a massive sewing kit, the random assortment of colors amounting to nothing. But if you looked at it through the glass sphere hanging from the wall, the spools morphed into a portrait of Leonardo da Vinci, appearing as if by magic.

Amid the assortment of space artifacts, the piece seemed out of place, as if the curators got confused and hung an impressionist painting in the Air and Space Museum—Degas's ballet dancers hanging next to the F-1 engine hardware. But here in Bezos's wonderland, where Dr. Seuss's quote was painted on the wall—"If you want to catch beasts you don't see every day, you have to go places quite out of the way"—it made sense. Going to space required looking through the prism to see that which was otherwise invisible.

IN A CONFERENCE room called Jupiter 2, Bezos settled into a chair with a cup of black coffee and nibbled from a small bowl of assorted nuts.

After years of secrecy, Blue Origin was starting to finally open up, and had even invited a small group of reporters to the headquarters a year before. But Bezos rarely granted one-on-one interviews like this, even to the *Washington Post*, the newspaper he had bought in 2013. It had taken me months of persistent cajoling to get this meeting.

The key, it seemed, was a press release from 1961 I'd discovered in the archives, highlighting the service of his grandfather, Lawrence P. Gise, as he left the Advanced Research Projects Agency to go back to the Atomic Energy Commission. Buttonholing Bezos after an event and handing him the press release was a last-ditch effort to get him to sit down for an interview. His grandfather was an important figure to him. Perhaps I could win him over with my level of research.

The world had seen him almost exclusively through the lens of Amazon, but to truly understand him, it also had to see him through another of his real passions: space. This was an important moment in the history of human space travel, one that needed to be more thoroughly chronicled. Would he be willing to sit down and talk about his ambitions building rockets?

He studied the release, looked at the photo of his grandfather, and listened to the pitch.

"I'm inclined to participate," he said, carefully.

It took another couple of months for "inclined" to solidify into a yes.

When Martin Baron, the executive editor of the *Post*, interviewed Bezos at a conference hosted by the paper, he acknowledged the tricky and potentially perilous position he was in. "In journalism, interviewing the owner of the company is considered to be high-risk behavior," he said.

As a *Washington Post* reporter, the same was true for me now.

Relaxed and in a good mood, sitting in the Jupiter 2 conference room, Bezos began to talk about his long-held passion for space, and what he hoped to accomplish. Amazon was still very much his primary occupation—something he was passionate about, especially as it had branched out from selling books to selling nearly everything. But on Wednesdays, he stole away to Blue's headquarters in Kent, Washington, about 20 miles south of downtown Seattle. Wednesdays were for space.

His girlfriend from high school had once told an interviewer that Bezos had founded Amazon in order to make enough money to start a space company. Now on this Wednesday in May 2017, he conceded that there "is some truth to that." The great fortune, which now stood at more than $80 billion, had allowed him to found Blue Origin.

While Musk had initially invested $100 million of his own money into SpaceX, the company had also benefited from more than $4 billion in contracts from NASA. By contrast, Bezos was his own NASA, funding Blue Origin almost entirely on his own. He'd joked that Blue Origin's business model has been, "I sell about $1 billion a year of Amazon stock, and I use it to invest in Blue Origin." He bought the *Washington Post* for $250 million in 2013. By contrast, he spent $2.5 billion of his own money on the New Glenn rocket alone, without accepting any government investment.

But Amazon was a real passion, he said, and no mere "stepping-stone" for Blue.

Bezos had recently been to the Oscars where Amazon Studios' film *Manchester by the Sea* had won an Academy Award. Alexa, Amazon's personal home assistant, was a hit, and the company was getting deeper into artificial intelligence. And groceries. Soon Amazon would acquire Whole Foods. There was plenty there to keep him occupied at what he called his "day job."

"I've fallen in love with it," he said.

He'd fallen in love with Blue, too. If you asked Alexa what she thought of Donald Trump, for example, she'd respond by saying, "When it comes to politics, I like to think big. We should be funding deep space exploration. I'd love to answer questions from Mars." And he had also recently scored a small role playing an alien in the film *Star Trek Beyond*.

A few days earlier, Bezos had been to the Seattle Museum of Flight, where the F-1, Apollo-era engines he had recovered from the bottom of the Atlantic Ocean had just gone on display. He had come to speak to a group of schoolchildren about the mission, and his interest in the cosmos, how from a young age he was "passionate about space, rockets, rocket engines, space travel."

"We all have passions," he told the students sitting before him on the floor. "You don't get to choose them, they pick you. But you have to be alert to them. You have to be looking for them. And when you find your passion, it's a fantastic gift for you because it gives you direction. It gives you purpose. You can have a job. You can have a career. Or you can have a calling."

His calling had come to him when he was five years old, as he watched Neil Armstrong take the first steps on the moon.

Sitting in the conference room, he said "I have a very distinct memory," he said, of the lunar mission and the moment space took hold of him. His grandparents and mother were huddled around the television. "I remember the excitement in the room," he continued. "And I remember the black-and-white TV."

Most of all, he remembered the sense "that something important was happening."

Something important was happening at Blue as well. Not far from where Bezos was seated, crews on the factory floor were building the next generation of New Shepard rockets, the ones that could take humans to space.

Over the past year, Blue had flown the same New Shepard booster five times in a row, with minimal refurbishment in between flights, pulling off precise landings each time and proving that reusability was possible. After each flight, the company painted a tortoise on the booster, a reminder of the slow, deliberate path. And the company had a new motto: "Launch. Land. Repeat."

The next step would be flying people on suborbital trips just past the edge of space. First would come test passengers—not pilots; the rocket flew itself autonomously—whose sole job would be to rate the experience from a customer's point of view. Were the seats comfortable? How was the view? Were the handles in the right place? Then would come the first space tourists, himself included.

"My singular focus is people in space," Bezos said. "I want people in space."

As a young child, he had wanted to be an astronaut. But as he got older, and studied rocketry, he realized that he wanted to also be an engineer. Armstrong was a hero, but he also admired Wernher von Braun, the German-born chief architect of the Apollo-era Saturn V rocket.

"I think he would also be quite disappointed that we aren't further out into space," Bezos said in response to a question about what von Braun would think of the state of the current space program. "I think he'd be shocked that nobody had been back to the moon. I think he'd be shocked that the record for the maximum number of humans in space at any one time is thirteen. He would be, like, 'What have you guys been up to? What, I die and the whole thing stops? *Dudes*, get on with it!'"

YEARS AFTER VIRGIN Galactic started touting its space tourism program, it now was about to have competition. Richard Branson was promising all the luxuries that had been associated with the glamorous Virgin brand. But with Amazon, Bezos had a long history of customer service experience that he was bringing to Blue.

Two days before the launch, Blue's passengers would arrive in West Texas, the company said on its website, where "the area's isolation lends clarity and focus as you prepare for the experience of a lifetime." A day-long training session would include an overview of the rocket and spacecraft, safety briefings, a simulation of the mission, and "maneuvering in a weightless environment"—"everything you need to know to make the most of your experience as an astronaut."

On the morning of the flight, as many as six passengers would board the spacecraft thirty minutes before launch time. Inside the capsule, with its white padded walls, they'd strap into reclined, La-Z-Boy–like seats each positioned next to a massive window, the biggest ever to fly into space.

They'd blast off in a burst of fire and smoke; soon the capsule would separate from the booster and float past the edge of space. The passengers would then unbuckle their seatbelts and have four minutes to float around the cabin, weightless. Thrusters would slowly spin the capsule to give passengers a 360-degree view. Then they'd strap themselves back into the seats and the spacecraft would fall back toward Earth under parachutes, before touching down in the desert. In all, the entire trip would last ten or eleven minutes.

No one sold space like Branson did, promising his customers a once-in-a-lifetime experience. But now Blue was getting into marketing mode, touting its program in similarly lofty tones.

"When you first look out through these massive windows, you just lose yourself in the panorama of blue and black," former NASA astronaut Nicholas Patrick, who worked as Blue's human integration architect, said in a promotional video the company posted on its website. "You can see clearly for millions of light years in every direction. It gives you a sense of the scale of the universe. The minute you unstrap, you're free. It opens up the possibilities for movement that you've just never, ever had here on Earth. It's a shared experience with your crew, but it's also profoundly personal. So, you really feel a part of the unfathomable depths of the cosmos."

And a connection to Earth. At least that's what the astronauts always said—that they went to space only to discover home. The crew of Apollo 8 had made it all the way to the far side of the moon, and then, as they came around the bend, there it was, "the pale blue dot," half-lit on the horizon, a frail planet, suspended in the darkness, alone. Their "Earth rise" photograph would become one of the most iconic images in the history of still photography.

In mid-2017, Bezos invited several of the surviving Apollo-era astronauts to an air show in Oshkosh, Wisconsin, where he was showing off the New Shepard booster and a mock-up of the crew capsule that would soon take paying tourists to space. It was an extraordinary reunion of the most exclusive of fraternities. There was Buzz Aldrin. And James Lovell, and Frank Borman, both members of the Apollo 8 crew, along with Fred Haise, who had served with Lovell on Apollo 13. There were Walter Cunningham, Apollo 7, and Gene Kranz, NASA's legendary flight director.

One by one, they stepped into Bezos's spacecraft, crossing the chasm from Apollo to the promise of the Next Giant Leap, even if so many of their brethren had passed away and would never see it come to fruition. They stretched out in the reclined seats next to the giant windows of this newly designed crew capsule, and ran their fingers over the handrails there to provide stability during those minutes of weightlessness. Bezos was thrilled. These were his heroes.

"Space changes people," he said, welcoming them. "Every time you talk to an astronaut, somebody who has been into space, they will tell you that when you look back at the Earth and see how beautiful it is and how fragile it is and that thin layer of the Earth's atmosphere that it makes you really appreciate home."

No one knew that more than the men assembled in his spacecraft.

"That was all heart," Bezos later said of the gathering. "I felt so many emotions, and had three of my four kids there."

Space had been a dream for decades, and he was looking forward to experiencing weightlessness, seeing the curvature of Earth, the blackness of space.

"I will go. I definitely will go. I can't wait, actually."

He had said that in 2007 on *Charlie Rose*. But now he was finally getting close to his dream.

FOR YEARS, BLUE Origin had been obsessively secretive, the tortoise holed up in his protective shell, not wanting to attract attention to himself, content to let the hare steal the show.

"We'll talk about Blue Origin when we have something to talk about," Bezos said at one point.

Now it had something to talk about—and had begun opening up ever so slightly. Over the course of several months, starting in early 2016, as New Shepard flew and flew again and as Bezos collected a series of awards on behalf of Blue Origin's groundbreaking landings, he made it clear in speeches and interviews that his ambitions went far beyond simply flying tourists to the edge of space and back.

After spending so many years researching and testing, he told a small group of reporters that he invited to Blue for a first ever media tour, "really exciting cool stuff that's not just hype is coming out the other end."

Without mentioning Musk by name, he said that "space is really easy to overhype." He continued, "There are very few things in the world where the ratio of attention you get to what you've actually done, can be extreme."

When asked about Musk, he said they "are very like-minded about a lot of things. We're not twins about our conceptualization about the future."

Bezos wanted to go to Mars, yes, but also "everywhere else." Musk liked to call Mars a "fixer-upper of a planet," one that he said could be heated up and made habitable in the event that an asteroid hit Earth and threatened to wipe out humanity.

Bezos seemed skeptical that Mars could be a backup for the human race. "[To] my friends who say they want to move to Mars one day, I say, 'Why don't you go live in Antarctica first for three years, and then see what you think?'" he said at a conference at the *Washington Post*. "'Because Antarctica is a garden paradise compared to Mars.'"

"Think about it," he said at another point. On Mars there was "no whiskey, no bacon, no swimming pools, no oceans, no hiking, no urban centers. Eventually Mars might be amazing. But that's a long way in the future."

NASA had visited every planet in the solar system, he would say, "and believe me, Earth is the best one. . . . This planet is incredible. There are waterfalls and beaches and palm trees and fantastic cities and restaurants and parties and events like this. And you're not going to get that anywhere but Earth for a really, really long time."

The better plan, then, was to preserve "this gem" called Earth. "We don't want Mars as a Plan B," he said. "Plan B is to make sure Plan A works. And Plan A is to make sure we keep this planet around for thousands of years."

It had become something of a well-honed stump speech, told over and over with remarkable consistency. Then again, using space as a way to preserve Earth was a concept he had been thinking about since high school.

"The whole idea is to preserve the Earth," he had told the *Miami Herald* in 1982 after his high school valedictorian speech. He was eighteen and saying Earth should be designated as a national park. Now, four decades later, he had revised his speech, only slightly. Instead of using the national park line, he said that Earth should be "zoned residential and light industrial."

The point was the same: all "heavy industry" would move into space. He now called this the "Great Inversion"—mining for energy resources in space, while leaving Earth alone. This planet was finite, Bezos said, lacking the resources to keep up with the demand of a world growing ever more developed and dynamic.

"There's all kinds of interesting stuff you can do around the solar system, but the thing that's going to move the needle for humanity the most is mining near-Earth objects and building manufacturing infrastructure in place," he said as he sat in the conference room at Blue Origin. "That's the big thing."

That would be in the distant future, after he was gone—"unless somebody does a good job on life extension," he added. But it was not that far in the future. "We only have a couple hundred years."

"If you take baseline energy usage today, compound it at just a few percent a year for just a few hundred years and you have to cover the entire Earth's surface with solar cells" to keep up with demand, he said. "You either go out into space or you need to control population on Earth. You need to control energy usage on Earth. These things are totally at odds with a free society. And it's going to be dull. I want my great-great-grandchildren to be using more energy per capita than I do. And the only way they can be using more energy per capita than me is if we expand out into the solar system. And then we can really keep Earth as this incredible gem that it is."

Blue Origin's oft-repeated goal was "millions of people living and working in space." But over the long term, it was even more ambitious than that. "If we want, we could have a trillion human beings living in the solar system," he said during an awards ceremony in Washington. "And then we'll have a thousand Einsteins, and a thousand Mozarts. What a cool civilization that would be."

When he started Amazon, the infrastructure was already in place so that a startup Internet company, even in 1995, could be successful. Now he wanted to start building the transportation network to space. While he had been inspired by the achievements during the Apollo era, the country's human spaceflight program had "been treading water for a long time," he said. And during an interview at the *Vanity Fair* "New Establishment Summit," he sounded very much like Musk, speaking about creating a "cargo route" to space that would be similar to the railroads that opened up the West.

"What I want to achieve with Blue Origin is to build the heavy-lifting infrastructure that allows for the kind of dynamic, entrepreneurial explosion of thousands of companies in space that I have witnessed over the last twenty-one years on the Internet," he said.

Amazon had its path laid out for it. Cables for the Internet had been laid. The postal service delivered packages to his customers. "There was already a payment system; we didn't have to do that," he said. "It was called the credit card, and it had been initially put in place for travelers."

All Amazon had to do then was "take that infrastructure and kind of reassemble it in a new way, and do something new and inventive with it. . . . In space today, that is impossible. On the Internet today, two kids in their dorm room can reinvent an industry, because the heavy-lifting infrastructure is in place for that. Two kids in their dorm room can't do anything interesting in space."

He wanted, then, to use his vast fortune to lay the foundation of that infrastructure into space. To make that part of his legacy.

"If I'm 80 years old and I'm looking back on my life," he said during an awards ceremony, "and I can say that I put in place, with the help of the teammates at Blue Origin, the heavy-lifting infrastructure that made access to space cheap and inexpensive so that the next generation could have the entrepreneurial explosion like I saw on the Internet, I'll be a very happy 80-year old."

BUT FIRST A relatively small step. Blue Origin would have to get good at launching reliably, efficiently, affordably, over and over, so that the act of going to space would became routine. Although suborbital space tourism had been derided by some as trivial, like bungee jumping in reverse for the superrich, as one science fiction author said, Bezos saw it as vital. If nothing else, the flights would be good practice.

"We humans don't get great at things we do a dozen times a year," Bezos said during a Q & A in 2016. Launching rockets at such a rate was "just not enough to get great at it. You never want to get a surgeon [who operates] just a dozen times a year. If you need to have surgery, find somebody who does the operation 20 to 25 times a week. That's the right level of practice."

Space tourism, then, was not just a way for people, albeit wealthy people, to experience space, but it was a way to make space more accessible.

"Tourism often leads to new technologies," Bezos said at the *Washington Post* forum. "And then those new technologies often circle back around and get used in very important, utilitarian ways." Graphic Processing Units, or GPUs, for example, were invented for video games. But now they're being used for machine learning, he said.

In addition to the ten-minute jaunts to space, the future for Blue Origin involved a much larger, more ambitious rocket. Internally, it had been called "Very Big Brother," but now it had a more formal name: New Glenn, after John Glenn, the first American in orbit.

Compared to New Shepard, the new rocket would be a beast, with seven engines, capable of 3.85 million pounds of thrust, towering as high as 313 feet, almost as tall as the Saturn V.

Eleven days before John Glenn died at the age of ninety-five in late 2016, he wrote Bezos a letter, saying he was "deeply touched" that the rocket was named after him. In 1962, when he took his historic flight into orbit, "you were still two years from being born," Glenn wrote. When Glenn returned to space in 1998 on a space shuttle mission at age seventy-seven, he noted that Blue Origin wouldn't be founded for another two years, but "you were already driven by a vision of space travel accessible not only to highly trained pilots and engineers and scientists, but to all of us. . . .

"As the original Glenn, I can tell you I see the day coming when people will board spacecraft the same way millions of us now board jetliners. When that happens, it will be largely because of your epic achievements this year."

Coming just days before the death of an American icon, the letter served as a bridge from the halcyon days of NASA's manned space program, Glenn's era of Mercury, Gemini, and Apollo, to this new era, a time Bezos had started to call a new "golden age of space exploration."

New Glenn, the smallest orbital rocket he'd ever build, would be capable of not just flying satellites and humans to low Earth orbit, but beyond. In Florida, Blue was building a massive manufacturing facility where it would build New Glenn. It was also revamping Pad 36, the launch complex just down the road from SpaceX's 39A. Over the past year, the company had gone on a hiring spree, and now had about one thousand employees.

Even though New Glenn was still at least three years from flying, in early 2017 Bezos announced that Blue had signed its first customer for the rocket, Eutelsat, a French satellite company. The deal would give Blue something that had been scarce in its history—actual revenue—and it marked the company's entrance into the market, where it would now compete against SpaceX.

NEW GLENN WAS yet another demonstration in the step-by-step approach. First came New Shepard, named after the first American in space. That took about ten years to develop. Then, New Glenn, which by the time it was scheduled to fly in 2020, would mark the culmination of another decade of work.

"We get to do a major thing every ten years," Bezos said, sitting in the conference room at Blue Origin's headquarters. "I think before I'm eighty, we have time for two more major cycles, maybe even two and a half. And so what those things will be I don't need to decide that now. It's premature. But if I can stay healthy, I'd like to see it. I'll make sure somebody will continue the work even if I'm not around to see it. I'd love to see it. I'm very curious about the future."

Working just one day a week at Blue meant time was precious. He stood up and headed out to his next meeting. Wednesdays were for space.

"And now I'm going to return to building rockets!" he said, as he walked out through the lobby.

It was hard to know what the future would look like, hundreds of years out. But he had big plans for how he'd fulfill the dreams of his five-year-old self. And he had recently given a very big hint about where he wanted to go.

His next rocket would be named New Armstrong.

EPILOGUE

Again, the Moon

P AUL ALLEN COULDN'T stay away.

After SpaceShipOne had made history as the first commercial vehicle to reach the edge of space, he had licensed the technology to Richard Branson, unnerved by the danger of the endeavor and ready to turn his attention, and fortune, elsewhere.

But space, and aviation, had been passions ever since he was a kid, and in 2011, he announced he was going to build the world's largest airplane. With a wingspan wider than a football field—end zones included—it would be larger than even the Spruce Goose, the famous aircraft Howard Hughes built during World War II that was designed to carry as many as seven hundred soldiers, but only flew once, in 1947.

Allen's plane wasn't designed to carry passengers; rather, rockets that would drop from the plane's belly at 35,000 feet and then launch into space. Because of its size, it would be capable of carrying rockets far more powerful than SpaceShipOne as well as carrying satellites, experiments, and eventually astronauts into orbit—not just to the threshold of space.

With the Ansari X Prize, Allen had been at the vanguard of the commercial space movement, which was now dominated by his fellow billionaire tycoons—Elon Musk, Jeff Bezos, and Branson, all of whom were pushing ahead with their own plans, showing it could be done. Allen wanted back in the game.

"You have a certain number of dreams in your life you want to fulfill," he said at the time. "And this is a dream that I'm very excited about."

His announcement came shortly after the space shuttle had flown its last mission and NASA was suddenly unable to fly astronauts into space. Despite the progress made by SpaceX and others, it was an uncertain time for the future of human space travel, and he noted, "with government-funded spaceflight diminishing, there is a much expanded opportunity for privately funded efforts." His new venture would, he said, keep "America at the forefront of space exploration."

Five years after that announcement, the plane was not yet ready to fly. But it was taking shape. Burt Rutan had retired from Scaled Composites, but Allen had hired his old company to build Stratolaunch in a massive hangar at the Mojave Air and Space Port which was so big the company had to apply for special construction permits just for the scaffolding.

Sitting in his Seattle office in August 2017, with views of the harbor, with the Seahawks' Super Bowl trophy nearby, he said that the plane was getting close to flying. Even unfinished, it was a behemoth. The wingspan seemed as long as a runway, and at 385 feet was longer than the distance traveled by the Wright Brothers on their first powered flights at Kitty Hawk. Its landing gear had a total of twenty-eight wheels. It had twin fuselages, and fully loaded it would weigh as much as 1.3 million pounds, be powered by six 747 engines, and have 60 miles of cable coursing through it.

In the history of aviation, there had never been anything quite like it. In addition to his fascination with space, Allen was a connoisseur of antique planes, and had amassed a collection of World War II relics that he had painstakingly refurbished. He recovered them from old battlefields—a Messerschmitt, a German fighter plane, was dug out of a sand dune on a French beach where it had been buried for decades; an Ilyushin IL-2M3 Sturmovik was pieced together from the wrecks of four planes recovered in northwest Russia.

To showcase his collection, Allen created a museum, the Flying Heritage & Combat Armor Museum in Everett, Washington, which featured a Grumman F6F-5 Hellcat, and a B-25 Bomber, among others.

"I would go in the university stacks and pull out books like *Jane's Fighting Aircraft of World War II* when I was 12 or something, and I'd spend hours reading about the engines in some of those planes," Allen recalled. "I was trying to understand how things worked—how things were put together, everything from airplane engines to rockets and nuclear power plants. I was just intrigued by the complexity and the power and the grace of these things flying."

Now he was building a plane as powerful and complex as any of them, built for opening up the cosmos. Allen's vision was like that of his fellow Space Barons—to lower the cost of space travel and make it more accessible. Bezos had said that inexpensive, reliable access to space would touch off the kind of "dynamic, entrepreneurial explosion of thousands of companies in space that I have witnessed over the last twenty-one years on the Internet."

Allen also saw parallels between the space frontier and the Internet.

"When such access to space is routine, innovation will accelerate in ways beyond what we can currently imagine," he said. "That's the thing about new platforms: when they become easily available, convenient, and affordable, they attract and enable other visionaries and entrepreneurs to realize more new concepts. . . .

"Thirty years ago, the PC revolution put computing power into the hands of millions and unlocked incalculable human potential. Twenty years ago, the advent of the web and the subsequent proliferation of smartphones combined to enable billions of people to surmount the traditional limitations of geography and commerce. Today, expanding access to LEO [low Earth orbit] holds similar revolutionary potential."

Just as computers had gone from the size of refrigerators to being able to fit in your pocket, once-massive and expensive satellites were smaller and cheaper, some even the size of a shoebox. Being able to put up constellations of thousands of them at a time would allow for all sorts of endeavors, from beaming the Internet to every spot in the world, opening up communication, to better monitoring the health of the planet, allowing farmers to keep a close eye on crops—and the Pentagon to monitor its enemies.

"The capabilities of these small satellites is something that's really interesting and fascinating, both for communications, where a lot of people are putting up constellations of satellites and for monitoring the challenged health of our planet," he said, sitting in his office. He'd become particularly interested in how space could be used to keep an eye on "things like illegal fishing in the ocean, which is an increasing problem."

In mid-2017, Heather Wilson, the new US Air Force secretary, visited the company's hangar in Mojave to discuss how it could be used to launch satellites for national security. Stratolaunch, able to take off and land at airports, could be a key player in launching those satellites quickly and affordably, as space was quickly becoming the next frontier in war.

While the X Prize flights had terrified him, Allen had begun to think about human space travel again. "I had to think long and hard about taking the plunge again," he wrote in his memoir. "Over time, my interest began to outweigh my reservations."

"Most exciting, for me, was the prospect of putting people into space for days and weeks at a time," Allen wrote. "I'd been happy to leave suborbital, high-volume space tourism to Richard Branson and Virgin Galactic. But there was something incomparably thrilling about orbital flight, going back to John Glenn's ride on the Friendship 7. It's an experience that goes way beyond a six-minute suborbital flight."

Richard Branson was building SpaceShipTwo for suborbital trips, but the company had been discussing the development of a more powerful rocket capable of sending humans into orbit. By 2017, Branson and Allen started discussing the possibility of launching that rocket off Stratolaunch in what would mark an extraordinary reunion.

The talks were preliminary, but "we hope we can work together on it," Branson said. "It would be quite nice, actually, since we started together, if we could end up working together again."

Allen didn't rule that out. But he also had plans of his own. In addition to creating a more reliable and efficient way to launch satellites, he was thinking bigger. Stratolaunch was so massive it could carry not just one rocket at a time but three, clustered under its belly like missiles on a fighter jet. But even three rockets wouldn't get close to the plane's capacity. He was also thinking about a reusable space shuttle called Black Ice that would be capable of flying to the International Space Station, taking satellites and experiments to orbit, and maybe one day, even people.

The ultimate goal was to have "airline-style operations," but for space, said Jean Floyd, Stratolaunch's chief executive officer. "You make your rocket a plane," he said. "So, you have an airplane carrying a plane that's fully reusable. You don't throw anything away ever. Only fuel."

A spaceplane—capable of not just delivering satellites to orbit, but of staying up for at least three days—that could be launched from virtually anywhere in the world. It was still in the development phase, a risky, push-the-envelope theory that might not pan out.

"I would love to see us have a full reusable system and have weekly, if not more often, airport-style, repeatable operations going," Allen said, while sitting in his Seattle office.

Returning to human spaceflight was a possibility sometime in the future, he said. "If you caught the bug back in the Mercury era, of course it's in the back of your mind. But I think you're seeing right now, other than [space station] resupply missions, most spaceflights are about launching satellites. That's the reality. And they are extremely important for everything from television to data all over the world. You can get data in the Kalahari desert because there's a satellite up there."

VIRGIN, MEANWHILE, HAD been taking out its new SpaceShipTwo, dubbed Unity, for test flights. Again and again, the mothership, WhiteKnightTwo, would hoist the spaceplane aloft, dropping it high above the Mojave Desert floor. Each test pushed the envelope further, until the company was finally getting close to the point where the testing program had been in 2014, when its spacecraft had come apart in midair.

As the testing progressed, Branson played the refrain he'd been singing for years: first flights were just around the corner. Always, just around the corner. After more than a decade of waiting to fly, Branson was nearing seventy, and getting itchy—as were his customers.

"I'm getting on, so we're going to have to hurry up," he said.

Now he had competition in Bezos's Blue Origin, which he relished. The space tourism experiences would be markedly different—Virgin Galactic's spaceplane versus Blue Origin's more traditional rocket.

"My guess is that quite a lot of people will want to try one and then try the other," he said. "And it's going to be interesting to see which passenger experience people enjoy the most."

He made clear, though, who he thought had the advantage: "We believe that going into space in a spaceship and coming back in that spaceship, on wheels, will be a customer experience that people would prefer than perhaps one or two other options that are being considered. And we'd love to see whether we're correct about that."

IN FEBRUARY 2017, SpaceX bounced back from its explosion with a mix of aplomb and audacity. It christened historic Launch Pad 39A for the first time since the last of the shuttle missions, resurrecting the once dormant site with a fiery flight of the Falcon 9 on a cargo mission to the space station.

A month earlier, the company had announced that it had found the cause of the explosion: not a rifle shot but a problem with a pressure vessel in the second-stage liquid oxygen tank. The tank had buckled, the company reported, and supercooled liquid oxygen propellant had pooled in the lining. The fuel had been ignited by breaking fibers or friction.

The Federal Aviation Administration had ruled out sabotage as a cause, and granted SpaceX a launch license. Musk concluded that "it was a self-inflicted wound. It took us a long time but we were able to re-create the failure. But it did alert us to the fact that sabotage was a real thing, so we upgraded the security." (A few months later, when a crew from CBS's *The Late Show with Stephen Colbert*, which was being escorted by Boeing officials, stopped outside the gates to check out SpaceX's use of 39A, SpaceX called security on them. They were stopped, questioned, and had to show identification before being allowed to go.)

Without proof of foul play in the rocket explosion, SpaceX pressed ahead, confident that it could endure another failure—even if the pair of explosions had been a blow to the company's finances and reputation.

"We've got cash in the bank, and we've got no debt," Gwynne Shotwell said at a press conference at the time. "So financially we're fine. It's hard to make money, though, in a year when you have a failure. So, I'm not going to kid anybody to say that wasn't a painful financial year for us last year, and frankly 2015. But it doesn't mean we're not a healthy and a vibrant company. We could withstand another failure for sure. I would not have done my job properly had we not been prepared for that."

Nothing cemented its status as the leader of the rising new industry than when the Falcon 9 lifted off from the same hallowed ground as the Apollo-era Saturn V—a launch that Musk called "an incredible honor." Shortly after nine a.m., the rocket rose with a thunderous, bone-rattling roar and then disappeared into a veil of low and dense clouds. Ten minutes later, however, it reappeared, as it flew back toward the landing pad, where it touched down softly.

By now, booster landings had almost become routine for SpaceX. It had a growing collection of so-called flight-proven first stages, all of which had landed successfully either on the landing pad or on the droneship at sea. What SpaceX had not done, however, was to re-fly one of those used rocket boosters. Landings were a wonderful bit of performance art that got millions of clicks on YouTube. But from a business standpoint, they were meaningless unless the rockets could be flown again and again.

As much as 70 percent of the cost of launch was in the booster stage, as Musk liked to say. It housed the most expensive and important part of the Falcon 9—its nine engines.

The first flight of a previously flown booster came a month later on a launch also from 39A. After the launch, an emotional Musk called it "an incredible milestone in the history of space," one that SpaceX had been working toward for fifteen years. This, he said, would be what would ultimately lower the cost of spaceflight, perhaps by a factor of a hundred or more—"the key to opening up space, and becoming a spacefaring civilization, a multiplanetary species and having the future be incredibly exciting and inspiring."

As it recovered from its explosion and moved through 2017, SpaceX screamed ahead, full force, racing through its backlog of seventy missions, worth some $10 billion. With six thousand employees, it at one point flew back-to-back missions within forty-eight hours, as it gobbled up a larger share of the international launch market.

SpaceX was struggling, however, with its Falcon Heavy rocket. It was years behind schedule and Musk would admit that the heavy-lift rocket, which had a total of twenty-seven engines that all had to fire at once, was "way, way more difficult than we originally thought. We were pretty naive about that." And he warned that the first launch could end up in a fireball.

"I hope it makes it far enough away from the pad that it does not cause pad damage. I would consider even that a win, to be honest," he said. "Major pucker factor, really. There's no other way to describe it."

In the meantime, SpaceX was struggling to meet NASA's rigorous requirements for the Dragon spacecraft that would fly astronauts to the space station. At NASA, there was a feeling among some that all of Musk's Mars talk was a distraction when what he really needed to focus on was flying the agency's most precious cargo—humans—to the space station. The agency had taken a huge gamble by selecting SpaceX, and it wasn't about to let its astronauts get on the Falcon 9, which had blown up twice, unless it felt confident the rocket was safe.

Musk said that was SpaceX's top priority, and had pushed back its timeline for Mars to focus on flying crews to the station. But as if colonizing Mars weren't enough, he was also planning to expand the company's already outsize ambitions, and rewrite its future. In early 2017, he made a surprise announcement that it had added a new destination to its itinerary, one it had eschewed up until now: the moon.

The mission would take two private citizens on a tourist trip that would orbit the moon and "travel faster and further into the Solar System than any before them," Musk announced.

Musk refused to name the passengers or how much they would pay, but he said the mission would be another step in "exceeding the high-water mark that was set in 1969 with the Apollo program." They wouldn't land on the lunar surface, but the weeklong trip would mark the first time humans had left low Earth orbit in decades.

It wasn't as challenging as Mars, but a lunar mission was also exceedingly difficult—and ambitious, considering that the company still hadn't flown anyone. The trip would take the passengers well past the moon, some 300,000 miles away, on a circumlunar trajectory, where it would use the moon's gravity to slingshot it back home.

Like the takeoff, the return would be perilous. The spacecraft would be flying as much as 40 percent faster as it hit Earth's atmosphere than would a return trip from the space station. And it had a very narrow window to it, or else it would bounce off the atmosphere and skip off into space.

IN GUADALAJARA, MUSK had unveiled a behemoth of a rocket that was so ambitious and mind-bogglingly large that critics said it was detached from reality. Since then, he had done some editing, and he presented a revised plan in September 2017 to build a massive, but more reasonably sized, version of what he called the BFR, or Big Fucking Rocket.

But while its size had been scaled back, its ambitions had not. In addition to helping create a city on the Red Planet, the new BFR would be capable of helping create a base camp on the moon.

"It's 2017; we should have a lunar base by now," he said during a speech. "What the hell has been going on?"

In a surprise twist, he also said that the massive rocket and spaceship, which would have more pressurized passenger space than an Airbus A380 airplane, could also fly people anywhere across the globe in less than an hour. Traveling at a maximum speed of nearly 17,000 mph above Earth's atmosphere, a trip from New York to Shanghai, for example, would take thirty-nine minutes, he said. Los Angeles to New York could be done in twenty-five.

"If we're building this thing to go to the moon and Mars, why not go other places as well?" he asked.

The new system would be capable of flying astronauts and cargo on an array of missions, including to the International Space Station in low Earth orbit. It could also launch satellites, he said, all of which would allow it to effectively replace the Falcon 9, Falcon Heavy, and Dragon spacecraft. In other words, after disrupting the industry, SpaceX would now attempt to disrupt itself.

But he made it clear that Mars remained the ultimate goal. During his talk, a chart showed that SpaceX planned to fly two cargo missions to Mars by 2022, a very ambitious timeline.

"That's not a typo," he said, but allowed: "It is aspirational."

By 2024, he said the company could fly four more ships to Mars, two with about one hundred human passengers each, sleeping two or three to a cabin, and two more cargo-only ships.

SpaceX had proven itself again and again in a series of improbable feats. It had a string of successful launches. It had pulled off landings no one thought possible. It had competed, and won, against the Alliance. It had had its failures, but had bounced back each time with triumphant launches.

And now as the hare tore down the track, making yet another audacious prediction that blurred the line between reality and fantasy, something remarkable was happening, at least in some corners.

People were starting to believe.

As Musk announced his plans to go to the moon, Bezos had been secretly talking to NASA about a lunar mission of his own.

Blue Origin distributed a secret plan it called "Blue Moon" to the leadership of NASA, urging it to back an Amazon-like delivery service that would bring cargo and supplies in support of a "future human settlement" of the lunar surface.

"It is time for America to return to the Moon—this time to stay," Bezos told the *Washington Post* after it obtained a copy of the seven-page report. "A permanently inhabited lunar settlement is a difficult and worthy objective." Flights to the moon could begin by 2020, he said, but only in partnership with the space agency. But he was "ready to invest my own money alongside NASA to make it happen."

As President Obama had pointed NASA toward Mars, he said of the moon that "we have been there before." It was technically true; men had left "flags and footprints" on the moon. But they hadn't been there in the permanent way Bezos and others were now proposing.

Bezos planned on landing cargo in a series of missions at Shackleton Crater at the moon's south pole, where there was nearly continuous sunlight that could power the spacecraft's solar arrays. And in the shadow of the crater, scientists had made the huge discovery of water ice. Water is not only key to human survival, but the oxygen and hydrogen could be used as another resource—fuel. Making the moon, then, a giant gas station in space.

Orbiting Earth, the International Space Station held a permanent, if small, colony. Now the moon could, too. But it could be bigger, with room for several nations to set up camps side by side.

Robert Bigelow, who was building inflatable space habitats that could be used to orbit the moon, said that "Mars is premature at this time. The moon is not."

Bezos believed that as well.

"I think that if you go to the moon first, and make the moon your home, then you can get to Mars more easily," he said.

AND SO THE moon. Again, the moon.

The greatest achievement in the history of humankind, revisited. Only now, so much time had passed that the twelve Apollo astronauts who had walked on the lunar surface were dying off, one by one.

James Irwin, Apollo 15, was the first to go, in 1991.

Alan Shepard, Apollo 14, died seven years later.

Pete Conrad, Apollo 12, passed a year after that.

Then Neil Armstrong, Apollo 11.

Then Edgar Mitchell, Apollo 14.

In January 2017, Gene Cernan, Apollo 17, the last man to walk on the moon, died. As he departed the lunar surface, Cernan said that "we leave as we came, and God willing, as we shall return with peace and hope for all mankind." He predicted that the return would be followed with a next giant leap, to Mars, by the end of the twentieth century, if not sooner.

Now, it was nearly fifty years since the height of the Space Age. The Apollo astronauts had blazed a trail that no one followed, their prophecies left unfulfilled.

Here, though, was a new generation, one ready to resurrect the dreams of their childhood, replicate the feats of their heroes, inspire as they had been inspired.

Bezos was five when he watched Armstrong walk on the moon. Musk had not yet been born. But with their massive fortunes and ambition, they were reenacting the Cold War space race, a pair of Space Barons starring in the roles of nations, hoping to pick up where Apollo had left off more than a generation earlier. Their race to the stars was driven not by war or politics; rather, by money and ego and adventure, a chance to extend humanity out into space for good.

They had taken their mark and the starter had fired his gun. The hare burst forward, kicking up a plume of dust. *Head down. Plow through the line.* The tortoise plodded along, step by step, repeating quietly, *Slow is smooth and smooth is fast.*

The race had been years in the making, but it had only just begun. It would continue down a long and unforeseeable path, until the years turned to decades and the decades into generations, lasting long after the tortoise and the hare were gone. A race past even their own imaginations, deep into the cosmos, to a point in the beyond where there was no finish line.

ACKNOWLEDGMENTS

The four billionaires featured in this book—Elon Musk, Jeff Bezos, Richard Branson, and Paul Allen—all run multiple companies and have huge demands on their time. So, I'm grateful that all of them graciously agreed to sit down with me and share their stories and insights. I'm thankful also that they approved many of my requests to speak with executives from their companies or associates, all of which made the narrative immeasurably better.

As it turns out, one of the subjects of this book, Jeff Bezos, is also the owner of my employer, *The Washington Post*. Let me address that head on. It is, I admit, somewhat awkward writing a book about someone who could have you fired. But under the leadership of Executive Editor Marty Baron, the *Post* has made it clear that it covers Jeff's companies as it would any other. Jeff receives the same treatment in these pages—fair and unflinching, without fear or favor.

I was first hired at the *Post* as a news aide when I was twenty years old—and have worked there for the majority of my adult life, enough time to have its values imprinted on my DNA, and to meet some of the most remarkable journalists practicing the craft today. Marty was gracious in allowing me leave to write this book. So were Cameron Barr, Emilio Garcia-Ruiz, Tracy Grant, and David Cho.

Three of my editors, Lynda Robinson, Dan Beyers, and Kelly Johnson, read various drafts of the manuscript, and helped shaped the book as much as anyone. Their support was gracious and overwhelming, and I owe each of them a huge debt of gratitude. I'm also grateful to Del Quentin Wilber for his counsel, passion, and keen eye.

The companies profiled here have dedicated staffs of immensely patient communications professions, who withstood my queries with grace.

Thank you to John Taylor, James Gleeson, and Sean Pitt at SpaceX; to Drew Herdener at Amazon; to Caitlin Dietrich at Blue Origin; to Christine Choi and Will Pommerantz at Virgin Galactic; to Steve Lombardi and Jim Jeffries at Vulcan. Thank you also to Tabatha Thompson and Mike Curie at NASA.

Eric Stallmer and Tommy Sanford at the Commercial Spaceflight Federation, both faithful advocates for the industry, helped open doors and were generous with their time and expertise. Many people in the space community also helped educate me about matters of policy, politics, and space. I'm grateful to James Muncy, Lori Garver, David Weaver, George Whitesides, Bretton Alexander, Tim Hughes, Phil Larson, Mike French, Stu Witt, Brendan Curry, and Rich Leshner.

The research for this book began while I was at the *Post*, as I chronicled the daily tumult of the beginnings of a new industry. But I relied greatly on the excellent reporting of many colleagues in the space press corps, including Jeff Foust, Joel Achenbach, Eric Berger, Irene Klotz, Frank Mooring Jr., Loren Grush, Alan Boyle, Stephen Clark, Kenneth Chang, Miriam Kramer, and James Dean.

In addition to the dozens of interviews I conducted for this book, my research depended on many texts, a few of which merit specific mention: Ashlee Vance's *Elon Musk: Tesla, SpaceX, and the Quest for a Fantastic Future*, Brad Stone's *The Everything Store: Jeff Bezos and the Age of Amazon*, and Julian Guthrie's *How to Make a Spaceship: A Band of Renegades, an Epic Race, and the Birth of Private Spaceflight*.

While I was on leave from the *Post*, I was fortunate to find another home—the Wilson International Center for Scholars—which provided a much-needed space to write and reflect. I'm thankful for the support of Jane Harman and Robert Litwak, who made the experience possible.

Rafe Sagalyn, my agent, was a relentless and enthusiastic advocate for the project. At PublicAffairs I owe thanks to my editor, John Mahaney, who shepherded the book from conception to print. Thank you also to Iris Bass and Sandra Beris for their careful copyedits.

Throughout the sometimes grueling process of this endeavor, I was lucky to have the love and support of my parents and extended family. My amazing children, Annie, Harrison, and Piper, were constant sources of joy, and reminders of what was really important. I'm grateful, above all, to my wife, Heather, a steadfast supporter and insightful reader, who inspires me daily. Love you and love you.

NOTES

1. "A SILLY WAY TO DIE"

12 **"We need to get out of here":** This account of the crash is based on interviews with Jeff Bezos, Ty Holland, and Brewster County sheriff Ronny Dodson; news reports, such as Gail Diane Yovanovich, "Chopper Crashes with Amazon.com Exec on Board," *Alpine Avalanche*, March 13, 2003; and federal investigative reports, including from the Federal Aviation Administration and the National Transportation Safety Board.

14 **Nearly a decade after Bezos:** Saul Hansel, "Amazon Cuts Its Loss as Sales Increase," *New York Times*, July 23, 2003.

15 **"It turned out to be":** Paul Geneson, "Dynamic Paseno: Charles 'Cheater' Bella," *El Paso Plus*, September 2, 2009.

16 **The local game warden:** Daniel Perez, "Cheater Bella Can't Escape Stigma of '88 Jailbreak," *El Paso Times*, July 11, 1997.

16 **The morning of the prison break:** Joline Gutierrez Krueger, "NM Had Its Own Love-Fueled Prison Break," *Albuquerque Journal*, June 17, 2015.

16 **She was obese:** Interview with Charles Bella, "Passion and Adventure," *Texas Monthly*, March 1990.

16 **"Her boyfriend is slapping me":** Ibid.

18 **"People say that your life":** Alan Deutschman, "Inside the Mind of Jeff Bezos," *Fast Company Magazine*, August 1, 2004.

19 **Although he wouldn't say:** Mylene Mangalindan, "Buzz in West Texas Is About Jeff Bezos and His Launch Site," *Wall Street Journal*, November 10, 2006.

19 **The mysterious buyer:** Ibid.

20 **"I have not been real pushy":** Sandi Doughton, "Amazon CEO Gives Us Peek into Space Plans," *Seattle Times*, January 14, 2005.

20 **But then one Monday:** John Schwartz, "Add to Your Shopping Cart: A Trip to the Edge of Space," *New York Times*, January 18, 2005.

20 **Since its founding in 2000:** Brad Stone, "Bezos in Space," *Newsweek*, May 5, 2003.

20 **And one industry official:** "One Small Step for Space Tourism . . . ," *Economist*, December 16, 2004.

22 **Stephenson held a variety:** Neal Stephenson, http://www .nealstephenson.com/blue-origin.html.

26 **"It became obvious that":** Steve Connor, "Galaxy Quest," *Independent*, August 4, 2003.

26 **"Those guys wanted to sell":** Brad Stone, "Amazon Enters the Space Race," *Wired*, July 2003.

2. THE GAMBLE

27 **"Doesn't anybody play higher":** Much of the discussion about Beal's trips to Vegas relied on Michael Craig's *The Professor, the Banker, and the Suicide King: Inside the Richest Poker Game of All Time* (New York: Grand Central Publishing, 2006).

28 **"This was not a comfort zone":** http://www.pokerlistings.com /poker-s-greatest-all-time-whales-andy-beal.

28 **"It is remarkable that":** R. Daniel Mauldin, "A Generalization of Fermat's Last Theorem: The Beal Conjecture and Prize Problem," *Notice of the American Mathematical Society* 44, no. 11 (December 1997).

29 **"We're broke":** http://www.pokerlistings.com/poker-s-greatest-all -time-whales-andy-beal.

30 **"Bluebonnet was like":** Thomas L. Moore and Hugh J. McSpadden, "From Bombs to Rockets at McGregor, Texas," American Institute of Aeronautics and Astronautics, January 2009.

30 **This was the approach:** Craig, *The Professor, the Banker, and the Suicide King*, p. 88.

31 **"If everybody else is going broke":** Melinda Rice, "Man with a Mission: The Founder of Beal Bank Is Seriously Rich and Seriously Smart. Now He's Serious About Shooting for the Stars," *D Magazine*, February 2000.

32 **"I don't lose a lot of sleep":** Ibid.

32 **In early 2000, the company:** Beal Aerospace, news release, "Beal Aerospace Test Fires Engine for BA-2 Rocket," March 6, 2000.

33 **In October 2000, Beal:** Andrew Beal, press release, "Beal Aerospace Regrets to Announce That It Is Ceasing All Business Operations Effective October 23, 2000."

35 **"Our parents had no idea":** Tom Junod, "Elon Musk: Triumph of His Will," *Esquire*, November 14, 2012.

36 **"I thought the Internet":** Elon Musk, "The Future of Energy and Transport," Oxford Martin School, Oxford University, November 14, 2012.

36 **"Well, I don't think you'll be coming back":** Elon Musk, "Stanford University Entrepreneurial Thought Leaders" lecture, October 8, 2003.

36 **"The online financial payment system":** Ibid.

37 **Given the size of the rock:** https://www.youtube.com/watch?v =xaW4Ol3_M1o.

38 **"We were both interested":** Junod, "Elon Musk."

38 **"Because, of course":** Elon Musk, "Mars Pioneer Award" acceptance speech, 15th Annual International Mars Society Convention, August 4, 2012.

38 **"I just did not want Apollo":** Pat Morrison Q & A with Elon Musk, "Space Case," *Los Angeles Times*, August 1, 2012.

39 **As a winged spaceplane:** Elon Musk, Stanford lecture.

39 **Space was still the exclusive:** For more on SpaceX's early days, see Ashlee Vance, "Elon Musk: Tesla, SpaceX and the Quest for a Fantastic Future," *Ecco*, May 19, 2015.

41 **On March 14, 2002, Musk founded:** Ibid.

41 **At the dawn of the Space Age:** Launch data compiled by the consulting firm Bryce Space and Technology.

41 **"I would bet you 1,000-to-one":** Jennifer Reingold, "Hondas in Space," *Fast Company* magazine, October 5, 2005.

43 **"The history of launch vehicle":** Jeff Foust, "The Falcon and the Showman," *Space Review*, December 8, 2003.

43 **"We're very proud to debut":** Ibid.

3. "ANKLE BITER"

47 **The young company:** Greg Lamm, "Rocket Maker Loses $227M Deal," *Pugent Sound Business Journal*, July 4, 2004.

48 **Kistler was hurting:** Citizens Against Government Waste press release, "NASA Yanks Sole-Source Contract After GAO Protest," June 24, 2004.

51 **One top air force official:** Much of the lawsuits with Northrop is from Jonathan Karp and Andy Pasztor, "Can Defense Contractors

Police Their Rivals Without Conflicts?" *Wall Street Journal*, December 28, 2004.

51 **"We do everything":** Ibid.

52 **"Northrop wasn't expecting us":** Ibid.

52 **As a kid in South Africa:** Ashlee Vance, "Elon Musk: Tesla, SpaceX and the Quest for a Fantastic Future," *Ecco*, May 19, 2015, 40.

53 **"I've never heard":** Renae Merle, "U.S. Strips Boeing of Launches; $1 Billion Sanction over Data Stolen from Rival," *Washington Post*, July 25, 2003.

53 **So, SpaceX filed suit:** *Space Exploration Technologies Corporation v. The Boeing Company and Lockheed Martin Corporation*, US District Court, Central District of California, case number CV05-7533, October 19, 2005.

54 **Boeing was just as dismissive:** Leslie Wayne, "A Bold Plan to Go Where Men Have Gone Before," *New York Times*, February 5, 2006.

54 **The failures were so frequent:** Vance, *Elon Musk*, 124.

55 **"I tell folks":** Sandra Sanchez, "SpaceX: Blasting into the Future—A Waco Today Interview with Elon Musk," *Waco Tribune*, December 22, 2011.

55 **Early on, Musk pegged:** Megan Geuss, "Elon Musk Tells BBC He Thought Tesla, SpaceX 'Had a 10% Chance at Success,'" *Ars Technica*, January 13, 2016.

56 **This was a man who:** http://www.10000yearclock.net/learnmore .html.

4. "SOMEWHERE ELSE ENTIRELY"

59 **Eisenhower entered:** Official White House Transcript of President Eisenhower's Press and Radio Conference #123, https://www.eisen hower.archives.gov/research/online_documents/sputnik/10_9_57 .pdf.

59 **In a memo to the White House:** Memo from C. D. Jackson regarding Soviet satellite, October 8, 1957, https://www.eisenhower.archives .gov/research/online_documents/sputnik/10_8_57_Memo.pdf.

60 **senator Lyndon Johnson fretted:** Matthew Brzezinski, *Red Moon Rising: Sputnik and the Hidden Rivalries That Ignited the Space Age* (New York: Henry Holt, 2007), 173–175.

60 **"In the 1960s you could do":** Charles Piller, "Army of Extreme Thinkers," *Los Angeles Times*, August 14, 2003.

61 **As a young employee:** Atomic Energy Commission, Meeting No. 410, 10:30 a.m., Thursday, May 18, 1950.

61 **"So the agency was controversial":** Richard J. Barber Associates, Inc., "The Advanced Research Projects Agency: 1958–1974," December 1975, http://www.dtic.mil/docs/citations/ADA154363.

61 **In a message to his colleagues:** Department of Energy Archives, minutes of meetings, 1961, https://www.osti.gov/opennet/search-results.jsp?full-text=L.%20Gise%20ALOO&sort-by=RELV&order-by=DESC.

62 **Gise would continue to serve:** Mark Leibovich, *The New Imperialists: How Five Restless Kids Grew Up to Virtually Rule Your World* (Saddle River, NJ: Prentice Hall, 2002), 70.

62 **He paid his son-in-law's tuition:** Brad Stone, *The Everything Store: Jeff Bezos and the Age of Amazon* (Boston: Back Bay Books/Little, Brown, 2013), 142.

62 **Jackie got a job:** Ibid.

62 **"I've never been curious":** Joshua Quittner, "An Eye on the Future: Jeff Bezos Merely Wants Amazon.com to Be the Earth's Biggest Seller of Everything," *Time*, December 27, 1999.

62 **"It really was a seminal moment":** Bezos Expeditions, http://www.bezosexpeditions.com/updates.html.

62 **On the ranch:** Joshua Quittner and Chip Bayers, "The Inner Bezos," *Wired Magazine*, March 1, 1999.

63 **"We'd hitch up the Airstream":** Jeff Bezos, "We Are What We Choose," baccalaureate address, Princeton University, May 30, 2010, https://www.princeton.edu/news/2010/05/30/2010-baccalaureate-remarks.

65 **The visits to the library:** Academy of Achievement, Washington, DC.

65 **"And from that day forward":** Ibid.

66 **"Will you please get":** Ibid.

66 **The engines were massive:** https://www.nasa.gov/topics/history/features/f1_engine.html.

67 **"I think I occasionally worried":** Ibid.

67 **"The whole idea is to preserve":** Sandra Dibble, "Ex-Dropout Leads His Class," *Miami Herald*, June 20, 1983.

67 **"He said the future of mankind":** Quittner and Bayers, "The Inner Bezos."

67 **In 1974, the *New York Times*:** Walter Sullivan, "Proposal for Human Colonies in Space Is Hailed by Scientists as Feasible Now," *New York Times*, May 13, 1974, 1.

68 **O'Neill strove to make:** Papers of Gerard O'Neill at the Archive at the Smithsonian National Air and Space Museum Steven F. Udvar-Hazy Center, Dulles, Virginia.

69 **O'Neill would "encourage":** Ibid.

71 **"Did I hear her right?":** From an interview with Kevin Scott Polk. The anecdote is also mentioned in his book *Gaiome: Notes on Ecology, Space Travel and Becoming Cosmic Species* (Booklocker.com, Inc., 2007).

72 **The set, which the catalog:** Russian Space History Sale 6516 Property of the Industries, Cosmonauts, and Engineers of the Russian Space Program, Sotheby's, December 11, 1993.

72 **It was a relatively low-cost:** Douglas Martin, "Space Artifacts of Soviets Soar at $7 Million Auction," *New York Times*, December 12, 1993.

75 **Once that was in place:** Alan Boyle, "Where Does Jeff Bezos Foresee Putting Space Colonists? Inside O'Neill Cylinders," Geekwire, October 29, 2016, https://www.geekwire.com/2016/jeff-bezos-space-colonies-oneill/.

76 **He replied that he'd just:** Jeffrey Ressner, "10 Questions for Jeff Bezos," *Time*, July 24, 2005.

76 **On March 5, 2005:** http://www.museumofflight.org/aircraft/charon-test-vehicle.

5. "SPACESHIPONE, GOVERNMENTZERO"

81 **But unlike other air-launched:** Ed Bradley, "The New Space Race," *60 Minutes*, November 7, 2004.

82 **"That was a pretty wild ride":** The account of the SpaceShipOne flights during the Ansari X Prize comes in large part from *Black Sky: Winning the X Prize*, the 2005 Discovery Channel documentary about the contest.

83 **"He flat didn't fly":** Eric Adams, "The New Right Stuff," *Popular Science*, November 1, 2004.

85 **Upset with what he saw:** Andrew Pollack, "A Maverick's Agenda: Nonstop Global Flight and Tourists in Space," *New York Times*, December 9, 2003.

87 **Rutan acknowledged:** Adams, "The New Right Stuff."

88 **"See what you're up against":** Julian Guthrie, *How to Make a Spaceship: A Band of Renegades, an Epic Race, and the Birth of Private Spaceflight* (New York: Penguin, 2016), 339.

88 **"Yeah," Rutan concurred:** Paul Allen, *Idea Man* (New York: Portfolio/Penguin, 2011).

89 **Left unsaid:** Guthrie, *How to Make a Spaceship*, 341.

91 **During Binnie's first:** Ibid., 229.

92 **Allen would see the prize!:** Ibid., 235.

93 **But suddenly Siebold:** Guthrie, *How to Make a Spaceship*, 360–361.

95 **On the morning of the flight:** Andrew Torgan, "Making History with SpaceShipOne: Pilot Brian Binnie Recalls Historic Flight," Space.com, October 2, 2014.

96 **"If I was this anxious":** Allen, *Idea Man*, 240.

6. "SCREW IT, LET'S DO IT"

101 **"With no fuel tanks":** The account of the balloon ride is based largely on Richard Branson, *Losing My Virginity: How I Survived, Had Fun, and Made a Fortune Doing Business My Way* (New York: Crown Business, 2007), 241.

102 **"The next thing, we found ourselves":** Howell Raines, "2 Trans-Atlantic Balloonists Saved After Jump into Sea off Scotland," *New York Times*, July 4, 1987.

103 **Her first flight:** Eve Branson, *Mum's the Word: The High-Flying Adventures of Eve Branson* (Bloomington, IN: AuthorHouse, 2013).

103 **And also from Captain:** The Penguin Q&A: Richard Branson, https://www.penguin.co.uk/articles/in-conversation/the-penguin-q-a/2015/nov/06/sir-richard-branson/.

104 **"I went round":** "Entrepreneurship Rubs Off When Filling Your First Plane," https://www.virgin.com/richard-branson/entrepreneurship-rubs-when-filling-your-first-plane.

104 **He called Boeing:** Branson, *Losing My Virginity*, 191–192.

104 **"We needed to":** Matt White, "1987: First People to Cross Atlantic in Hot Air Balloon," Guinness Book of World Records, August 18, 2015, http://www.guinnessworldrecords.com/news/60at60/2015/8/1987-first-people-to-cross-the-atlantic-in-a-hot-air-balloon-392904.

104 **Carefully, he fired the burner:** Branson, *Losing my Virginity*, 247.

104 **As he recalled:** Michael Specter, "Branson's Luck," *New Yorker*, May 14, 2007.

105 **By 1977, when Branson:** Richard Branson, "I Found the Policeman Who Arrested Us for Selling Never Mind the Bollocks," https://www.virgin.com/richard-branson/i-found-the-policeman-who-arrested-us-for-selling-never-mind-the-bollocks.

105 **The first attempt:** Branson, *Losing My Virginity*, 217.

109 **Despite the daunting task:** Jill Lawless, "Space-Flight Tickets to Start at $208,000," Associated Press, September 28, 2004.

109 **"We hope to create":** "Now Virgin to Offer Trips to Space," CNN, September, 27, 2004, http://www.cnn.com/2004/WORLD/europe/09/27/branson.space/.

110 **"We like to think":** "200 on Pan Am Waiting List Are Aiming for Moon," *New York Times*, January 9, 1969.

110 **The list grew quickly:** Jeff Gates, "I Was a Card-Carrying Member of the 'First Moon Flights' Club," http://www.smithsonianmag.com/smithsonian-institution/i-was-card-carrying-member-first-moon-flights-club-180960817/.

110 **"Commercial flights to the moon":** Robert E. Dallos, "Pan Am Has 90,002 Reservations: Public Interest Grows in Flights to the Moon," *Los Angeles Times,* February 19, 1985.

112 **Branson's version of space:** Paul Allen, *Idea Man* (New York: Portfolio/Penguin, 2011), 243.

7. THE RISK

117 **"The United States is a distillation":** Elon Musk, "Mars Pioneer Award" acceptance speech, 15th Annual International Mars Society Convention, 2012.

119 **As a guidebook pointed out:** David Goodman, *Best Backcountry Skiing in the Northeast* (Boston: Appalachian Mountain Club Books, 2010).

120 **In modern society:** Paul O'Neil, *The Epic of Flight, Barnstormers & Speed Kings* (New York: Time-Life Books, 1981).

121 **"If we die":** John Barbour, "Footprints on the Moon," Associated Press, 1969.

121 **Gene Kranz, the flight director:** Nova online, interview with Gene Kranz, http://www.pbs.org/wgbh/nova/tothemoon/kranz .html.

122 **Musk had always had a bit:** Kerry A. Dolan, "How to Raise a Billionaire: An Interview with Elon Musk's Father, Errol Musk," *Forbes,* July 12, 2015.

122 **His maternal grandparents:** "Tesla and SpaceX: Elon Musk's Industrial Empire," Segment Extra, "Elon Musk on His Family History," *60 Minutes,* March 30, 2014.

122 **"There is something particularly":** Fay Goldie, *Lost City of the Kalahari: The Farini Story and Reports on Other Expeditions* (Cape Town: A. A. Balkema, 1963).

123 **Their guide slept:** Ibid.

123 **"The thing that actually":** Musk, "Mars Pioneer Award" acceptance speech.

124 **James Oberstar, a longtime:** "Commercial Space Transportation: Beyond the X Prize," hearing before the Subcommittee of Aviation of the Committee on Transportation and Infrastructure, US House of Representatives, 109th Congress, February 9, 2005.

8. A FOUR-LEAF CLOVER

128 **"To cast a javelin":** http://www.darpa.mil/about-us/mission.

129 **Over the years:** Robert M. Gates and the DARPA media staff, *DARPA: 50 Years of Bridging the Gap* (Washington, DC: Faircount LLC, 2008).

130 **The launch was supposed to:** Leonard David, "SpaceX Private Rocket Shifts to Island Launch," Space.com, August 12, 2005.

131 **"It's like you build":** Ibid.

133 **"Commercial enterprises":** NASA Johnson Space Center Oral History Project, Commercial Crew & Cargo Program Office, view by Rebecca Wright, January 12, 2013.

133 **NASA wanted to know:** NASA Oral History Project, January 15, 2013.

134 **Starting as soon as it received:** Ibid.

134 **"A million things":** "SpaceX Aims to Regain Momentum with New Rocket Launch," CBS News, January 13, 2017.

134 **"If we have three consecutive failures":** David, "SpaceX Private Rocket Shifts to Island Launch."

136 **Afterward, Musk tried:** Tariq Malik, "SpaceX's Inaugural Falcon 1 Rocket Lost Just After Launch," Space.com, March 24, 2006.

137 **"I think it could be some":** "NASA Awards Two Contracts to Develop Private Spaceship," Bloomberg News, August 19, 2006.

137 **"I'm tired of hearing that":** "NASA Picks 2 Firms for Private Spaceship," Associated Press, August 19, 2006.

138 **"It was new to everybody":** NASA Oral History Project, June 12, 2013.

138 **"Commercial companies":** NASA Oral History Project, March 1, 2013.

139 **"The funding is milestone-based":** Irene Klotz, "U.S. Rocket Firm Puts Malaysian Satellite into Orbit," Reuters, July 14, 2009.

140 **Looking back on it:** Michael Griffin, NASA Oral History Project, January 12, 2013.

141 **"I'm going to watch":** Carl Hoffman, "Elon Musk Is Betting His Fortune on a Mission Beyond Earth's Orbit," *Wired*, May 22, 2007.

141 **"This was a pretty nerve-racking":** Tariq Malik, "SpaceX's Second Falcon 1 Rocket Fails to Reach Orbit," Space.com, March 20, 2007.

141 **"SpaceX will not skip":** John Schwartz, "Launch of Private Rocket Fails; Three Satellites Were Onboard," *New York Times*, August 3, 2008.

141 **He added: "For my part":** Jeremy Hsu, "SpaceX's Falcon 1 Falters for a Third Time," Space.com, August 3, 2008.

141 **The challenge "was":** Gwynne Shotwell, NASA Oral History, January 15, 2003.

142 **"Between the third":** Hans Koenigsmann, NASA Oral History, January 15, 2013.

142 **"We wanted to keep":** https://www.nasa.gov/feature/the-making-of -the-apollo-11-mission-patch.

145 **And when the NASA officials:** "Tesla and SpaceX: Elon Musk's Industrial Empire," *60 Minutes*, March 30, 2014.

145 **Goddard was derided:** "Apollo 11: How America Won the Race to the Moon," Associated Press, August 21, 2016.

145 **"That Professor Goddard":** "A Severe Strain on Credulity," *New York Times*, January 13, 1920.

145 **Goddard responded by saying that:** https://www.nasa.gov/missions /research/f_goddard.html.

146 **"How many more years":** "Apollo 11."

146 **"Further investigation":** "A Correction," *New York Times*, July 17, 1969.

9. "DEPENDABLE OR A LITTLE NUTS?"

150 **Musk was thrilled:** https://www.youtube.com/watch?time_continue =11&v=CUmnzaDGifo.

153 **When the company was told:** Irene Klotz, "SpaceX Secret? Bash Bureaucracy, Simplify Technology," *Aviation Week & Space Technology*, June 15, 2009.

154 **When it was building Falcon 1:** Jennifer Reingold, "Hondas in Space," *Fast Company Magazine*, February 1, 2005.

154 **The rocket's avionics:** Ibid.

154 **Instead of using the straps:** John Couluris, NASA Johnson Space Center Oral History Project, Commercial Crew & Program Office, January 15, 2003.

155 **At "SpaceX, we weren't":** Ibid.

155 **"The biggest challenge":** Gwynne Shotwell, NASA Oral History, January 15, 2003.

156 **"When we talked to them":** Michael Horkachuck, NASA Oral History, November 6, 2012.

160 **"The president's proposed":** Joel Achenbach, "Obama Budget Proposal Scraps NASA's Back-to-the-Moon Program," *Washington Post*, February 2, 2010.

160 **Michael Griffin, the former:** Joel Achenbach, "NASA Budget for 2011 Eliminates Funds for Manned Lunar Missions," *Washington Post*, February 1, 2010.

164 **"I think he wanted":** Marc Kaufman, "One Giant Leap for Privatization?" *Washington Post*, June 4, 2010.

164 **"I hope people don't":** Marcia Dunn, "PayPal Millionaire's Rocket Making 1st Test Flight," *Associated Press*, June 3, 2010.

165 **"A dramatic launch failure":** Andy Pasztor, "Space Pioneer Elon Musk Faces Big Risks with Upcoming Launch," *Wall Street Journal*, June 4, 2010.

167 **Eventually a reporter:** Andy Pasztor, "Amazon Chief's Spaceship Misfires," *Wall Street Journal*, September 3, 2011.

10. "UNICORNS DANCING IN THE FLAME DUCT"

171 **"Miraculous profits await you":** Gary White, "Miracle City Mall Was Once a Bright Spot in Titusville," *Lakeland Ledger*, June 23, 2011.

174 **And a spokesman admitted:** Scott Powers, "NASA Picks SpaceX to Run KSC Launch Complex," *Orlando Sentinel*, December 13, 2013.

174 **Musk even put a price tag:** Jonathan Amos, "Mars for the 'Average Person,'" BBC News, March 20, 2012.

175 **At SpaceX's headquarters:** Brian Vastag, "SpaceX Dragon Capsule Docks with International Space Station," *Washington Post*, May 25, 2012.

175 **"This is, I think":** Kenneth Chang, "First Private Craft Docks with Space Station," *New York Times*, May 25, 2012.

182 **In a statement to *SpaceNews*:** Dan Leone, "Musk Calls Out Blue Origin, ULA for 'Phony Blocking Tactic' on Shuttle Pad Lease," *SpaceNews*, September 25, 2013.

182 **It enlisted the aid:** Alan Boyle, "Billionaires' Battle for Historic Launchpad Goes into Overtime," NBC News, September 18, 2013.

183 **"It is therefore unlikely":** Leone, "Musk Calls Out Blue Origin."

11. MAGIC SCULPTURE GARDEN

187 **Some 15 miles off the coast:** Martin Weil, "Storm Rips Apart Commercial Fishing Boat off Maryland's Coast," *Washington Post*, March 8, 2013.

189 **"The *Titanic* looks":** David Concannon, "Titanic: The First Dive of a New Century," *Fathoms Magazine*, no. 6.

190 **"It's hard to find":** Bezos Expeditions produced a video and published a series of updates on its website, http://www.bezosexpeditions.com/updates.html.

191 **The side-scan sonar system:** http://www.blacklaserlearning.com/adventure/how-do-you-recover-an-apollo-rocket-engine-from-over-2-miles-beneath-the-bermuda-triangle/.

192 **After studying the data:** Bezos Expeditions, http://www.bezosexpeditions.com.

193 **"You can feel it walking":** Ibid.

193 **"Three miles below":** Ibid.

194 **"Mariners all over the world":** Ibid.

196 **Bezos announced the news:** Ibid.

197 **On the menu:** Michael Y. Park, "Eating Maggots: The Explorers Club Dinner," *Epicurious,* March 17, 2008, http://www.epicurious .com/archive/blogs/editor/2008/03/eating-maggots.html.

197 **One year, the club's president:** Lynda Richardson, "Explorers Club: Less 'Egad' and More 'Wow!' *New York Times*, December 3, 2004.

197 **"Jeff is trying to get people":** https://archive.org/details/ECAD 2014720_201502.

12. "SPACE IS HARD"

203 **And it was only a test:** Christian Davenport, "SpaceX Rocket Blows Up over Texas," *Washington Post*, August 25, 2014.

204 **The *Atlantic* canonized him:** Nicole Allan, "Who Will Tomorrow's Historians Consider Today's Greatest Inventors," *Atlantic*, November 2013, https://www.theatlantic.com/magazine/archive/2013/11 /the-inventors/309534/.

205 **National security launches paid:** "The Air Force's Evolved Expendable Launch Vehicle Competitive Procurement," US Government Accountability Office, March 4, 2014, http://www.gao.gov/assets /670/661330.pdf.

205 **"Musk is":** Aaron Mehta, "Elon Musk on Russian Assassins, Lockheed Martin, and Going to Mars," *Defense News*, June 10, 2014, http://intercepts.defensenews.com/2014/06/elon-musk-on-russian -assassins-lockheed-martin-and-going-to-mars/.

206 **"Our toughest competitor":** Ibid.

207 **"SpaceX is trying to cut":** Christian Davenport, "ULA Chief Accuses Elon Musk's SpaceX of Trying to 'Cut Corners,'" *Washington Post*, June 18, 2014.

208 **"It's kind of the best":** Joel Achenbach, "Jeff Bezos' Blue Origin to Supply Rocket Engines for National Security Launches," *Washington Post,* September 17, 2014, embedded video, https://www .washingtonpost.com/national/health-science/jeff-bezos-and-blue -origin-to-supply-engines-for-national-security-space-launches /2014/09/17/59f46eb2-3e7b-11e4-9587-5dafd96295f0_story.html ?utm_term=.be88d6562a8d.

208 **"If all your competitors":** Andrea Shalal, "Boeing-Lockheed Venture Picks Bezos Engine for Future Rockets," Reuters, September 17, 2014.

209 **"That is how a 21st Century":** Tony Reichardt, "That Is How a 21st Century Spaceship Should Land," *Smithsonian Air & Space*, May 30, 2014, http://www.airspacemag.com/daily-planet/ii-how-21st -century-spaceship-should-land-180951621/.

210 **Leading up to the launch:** Marcia Dunn, "Space Station Supply Launch Called Off in Virginia," Associated Press, October 28, 2014.

211 **The first flights were supposed to start:** Virgin Galactic Overview, https://web.archive.org/web/20070331154530/http://virgingalactic .com/htmlsite/overview.htm.

211 **For $250,000, Virgin promised:** Ibid.

211 **It had inked a deal:** "NBCUniversal Announces Exclusive Partnership with Sir Richard Branson's Virgin Galactic to Televise First Commercial Flight to Space," press release, November 8, 2013.

212 **Flying above the Mojave Desert:** "G Force Training with Virgin Galactic," October 8, 2014, https://www.virgin.com/richard -branson/g-force-training-virgin-galactic.

212 **They were close friends:** Christian Davenport and Jöel Glenn Brenner, "Two Pilots Who Were Close Friends Now Tied Together by One Fatal Flight," *Washington Post*, November 3, 2014.

213 **Siebold had considered the mission:** The account of the crash comes from the National Transportation Safety Board's investigation, https://www.ntsb.gov/news/events/Pages/2015_spaceship2 _BMG.aspx.

218 **But a NASA slide showed:** Jeff Foust, "Progress Anomaly Strains Space Station Supply Lines," *SpaceNews*, April 28, 2015.

219 **"The vast majority of people":** Christian Davenport, "Hearing Elon Musk Explain Why His Rocket Just Blew Up Shows Why He's Such an Intense CEO," *Washington Post*, June 20, 2015.

13. "THE EAGLE HAS LANDED"

221 **The capsule on top of the rocket:** "Blue Origin Makes Historic Rocket Landing," November 24, 2015, https://www.blueorigin.com /news/news/blue-origin-makes-historic-rocket-landing.

222 **In interviews afterward:** Christian Davenport, "Jeff Bezos Sticks Rocket Landing, Stakes Claim in Billionaires' Space Race," *Washington Post*, November 24, 2015.

223 **"The pad has stood silent":** Christian Davenport, "Jeff Bezos's Blue Origin Space Company to Launch from Historic Pad at Space Coast," *Washington Post*, September 15, 2015.

224 **Reaching the threshold of space:** Christian Davenport, "The Inside Story of How Billionaires Are Racing to Take You to Outer Space," *Washington Post*, August 19, 2016.

225 **As Musk once said:** Carl Hoffman, "Elon Musk Is Betting His Fortune on a Mission Beyond Earth Orbit," *Wired*, May 22, 2007.

227 **SpaceX compared it to:** "X Marks the Spot: Falcon 9 Attempts Ocean Platform Landing," December 16, 2014, http://www.spacex

.com/news/2014/12/16/x-marks-spot-falcon-9-attempts-ocean
-platform-landing.

228 **"Well, at least we got close":** Christian Davenport, "After SpaceX
Sticks Its Landing, Elon Musk Talks About a City on Mars," *Washington Post*, December 22, 2015.

229 **"It really quite dramatically":** Ibid.

231 **The National Transportation Safety Board:** NTSB press release,
"Lack of Consideration for Human Factors Led to In-flight Breakup
of SpaceShipTwo," July 28, 2015.

231 **As board member Robert Sumwalt:** Christian Davenport, "NTSB
Blames Human Error, Compounded by Poor Safety Culture, in Virgin Galactic Crash, *Washington Post*, July 28, 2015.

234 **"At Blue Origin, our biggest":** Christian Davenport, "Jeff Bezos on
Nuclear Reactors in Space, the Lack of Bacon on Mars and Humanity's Destiny in the Solar System," *Washington Post*, September 15,
2016.

235 **"If somebody can just tell me":** John Logsdon, *John F. Kennedy and
the Race to the Moon* (New York: Palgrave Macmillan, 2010), 77–78.

14. MARS

238 **"Essentially what we're saying":** Christian Davenport, "Elon Musk
Provides New Details on His 'Mind Blowing' Mission to Mars,"
Washington Post, June 10, 2016.

239 **"So," he said, "how do we figure out":** "Making Humans a Multiplanetary Species," http://www.spacex.com/mars.

241 **"The priorities of all of our customers":** "United Launch Alliance Announces Rapid Launch, the Industry's Fastest Order to
Launch Service," September 13, 2016, http://www.ulalaunch.com
/ula-announces-rapidlaunch.aspx.

241 **A SpaceX employee suddenly:** Christian Davenport, "Implication
of Sabotage Adds Intrigue to SpaceX Investigation," *Washington
Post*, September 30, 2016.

243 **But it was also all a bit:** Christian Davenport, "Elon Musk on Mariachi Bands, Zero-G Games, and Why His Mars Plan Is Like 'Battlestar Galactica,'" *Washington Post*, September 28, 2016.

243 **If Musk were going to be able:** Davenport, "Implication."

244 **The $23 billion SLS/Orion program:** "NASA Human Space Exploration: Opportunity Nears to Reassess Launch Vehicle and Ground
Systems Cost and Schedule," US Government Accountability Office,
July 2016.

247 **"His job is to provide":** Christian Davenport, "Elon Musk Offers
Glimpse of Plans to Deliver Humans to Mars," *Washington Post*,
September 27, 2016.

248 **"There's so much interest":** Christian Davenport, "Why Investors Are Following Musk, Bezos in Betting on the Stars," *Washington Post*, January 28, 2016.

248 **By mid-2017, after raising $350 million:** Katie Benner and Kenneth Chang, "SpaceX Is Now One of the World's Most Valuable Privately Held Companies," *New York Times*, July 27, 2017.

249 **"We believe space mining":** Lauren Thomas, "In a New Space Age, Goldman Suggests Investors Make It Big in Asteroids," CNBC, April 6, 2017.

15. "THE GREAT INVERSION"

254 **He'd joked that Blue Origin's business model:** Christian Davenport, "Jeff Bezos Shows Off the Crew Capsule That Could Soon Take Tourists to Space," *Washington Post*, April 5, 2017.

254 **By contrast, he spent $2.5 billion:** Caleb Henry, "Blue Origin Enlarges New Glenn's Payload Fairing, Preparing to Debut Upgraded New Shepard," *SpaceNews*, September 17, 2017.

254 **"We all have passions":** Alan Boyle, "Video: Watch Amazon's Jeff Bezos Talk with Kids About Apollo's Space Legacy—and Share Life Lessons," Geekwire, May 20, 2017, https://www.geekwire.com/2017 /jeff-bezos-kids-apollo/.

256 **Two days before the launch:** https://www.blueorigin.com/astronaut -experience.

258 **"We'll talk about Blue Origin":** Davenport, "Why Jeff Bezos Is Finally Ready to Talk About Taking People to Space," *Washington Post*, March 8, 2016.

258 **Without mentioning Musk:** Ibid.

258 **"Think about it," he said:** Christian Davenport, "Jeff Bezos on Nuclear Reactors in Space, the Lack of Bacon on Mars, and Humanity's Destiny in the Solar System," *Washington Post*, September 15, 2016.

260 **While he had been inspired:** Calla Cofield, "Spaceflight Is Entering a New Golden Age, Says Blue Origin Founder Jeff Bezos," Space .com, November 25, 2015, https://www.space.com/31214-spaceflight -golden-age-jeff-bezos.html.

261 **"If I'm 80 years old":** Ibid.

261 **Although suborbital space tourism:** John Thornhill, "Mars Visionaries Herald a New Space Age," *Financial Times*, August 21, 2017.

261 **"We humans don't get great":** Alan Boyle, "Interview: Jeff Bezos Lays Out Blue Origin's Space Vision, from Tourism to Off-planet Heavy Industry," Geekwire, April 13, 2016.

261 **Eleven days before John Glenn:** Brian Wolly, "Read the Letter Written by John Glenn to Honor Jeff Bezos for Blue Origin," *Smithsonian Magazine*, December 8, 2016, http://www.smithsonianmag.com

/innovation/read-letter-written-sen-john-glenn-honor-jeff-bezos-blue-origin-180961366/.

262 **Coming just days before:** Cofield, "Spaceflight Is Entering a New Golden Age."

EPILOGUE: AGAIN, THE MOON

265 **"You have a certain number":** Kenneth Chang, "Tycoon's Next Big Bet for Space: A Countdown Six Miles Up in the Air," *New York Times*, December 13, 2011.

266 **In addition to his fascination with space:** http://www.flyingheritage.com/Explore/The-Collection/Russia/Ilyushin-II-2M3-Shturmovik.aspx.

266 **"I would go in the university stacks":** Clare O'Connor, "Inside Microsoft Mogul Paul Allen's Multi-Million Dollar WWII Plane Collection," *Forbes*, June 4, 2013.

267 **"When such access to space":** Christian Davenport, "Why Microsoft Co-founder Paul Allen Is Building the World's Largest Airplane," *Washington Post*, June 20, 2016.

271 **It was years behind schedule:** https://www.youtube.com/watch?v=sDNdYgh5124.

274 **Robert Bigelow:** Christian Davenport, "An Exclusive Look at Jeff Bezos's Plan to Set Up Amazon-like Delivery for 'Future Human Settlement' of the Moon", *Washington Post*, March 2, 2017.

INDEX

Copyright © Marvin Joseph, *Washington Post*

Christian Davenport has been a staff writer at *The Washington Post* since 2000 and currently covers the space and defense industries for the financial desk. He has also worked at *Newsday,* the *Philadelphia Inquirer,* and the *Austin American-Statesman.* He is a recipient of the Peabody award for his work on veterans with traumatic brain injury and has been on reporting teams that were finalists for the Pulitzer Prize three times. He is the author of *As You Were: To War and Back with the Black Hawk Battalion of the Virginia National Guard.* He lives in Washington, DC, with his wife and three children.

PublicAffairs is a publishing house founded in 1997. It is a tribute to the standards, values, and flair of three persons who have served as mentors to countless reporters, writers, editors, and book people of all kinds, including me.

I. F. STONE, proprietor of *I. F. Stone's Weekly*, combined a commitment to the First Amendment with entrepreneurial zeal and reporting skill and became one of the great independent journalists in American history. At the age of eighty, Izzy published *The Trial of Socrates*, which was a national bestseller. He wrote the book after he taught himself ancient Greek.

BENJAMIN C. BRADLEE was for nearly thirty years the charismatic editorial leader of *The Washington Post*. It was Ben who gave the *Post* the range and courage to pursue such historic issues as Watergate. He supported his reporters with a tenacity that made them fearless and it is no accident that so many became authors of influential, best-selling books.

ROBERT L. BERNSTEIN, the chief executive of Random House for more than a quarter century, guided one of the nation's premier publishing houses. Bob was personally responsible for many books of political dissent and argument that challenged tyranny around the globe. He is also the founder and longtime chair of Human Rights Watch, one of the most respected human rights organizations in the world.

· · ·

For fifty years, the banner of Public Affairs Press was carried by its owner Morris B. Schnapper, who published Gandhi, Nasser, Toynbee, Truman, and about 1,500 other authors. In 1983, Schnapper was described by *The Washington Post* as "a redoubtable gadfly." His legacy will endure in the books to come.

Peter Osnos, *Founder*